Springer Series in Optical Sciences

Volume 50

Edited by Theodor Tamir

Springer Series in Optical Sciences

Editorial Board: J.M. Enoch D.L. MacAdam A.L. Schawlow K. Shimoda T. Tamir

Volumes 1–41 are listed on the back inside cover

H. J. Eichler · P. Günter
D. W. Pohl

Laser-Induced Dynamic Gratings

With 123 Figures

Springer-Verlag Berlin Heidelberg GmbH

Professor Dr. HANS JOACHIM EICHLER
Technische Universität Berlin, Optisches Institut, Straße des 17. Juni 135
D-1000 Berlin 12

PD Dr. PETER GÜNTER
Laboratorium für Festkörperphysik, Eidgenössische Technische Hochschule
ETH-Hönggerberg, CH-8093 Zürich, Switzerland

Dr. DIETER W. POHL
IBM Zürich, Research Laboratory, CH-8803 Rüschlikon, Switzerland

ISBN 978-3-662-15197-6   ISBN 978-3-540-39662-8 (eBook)
DOI 10.1007/978-3-540-39662-8

Preface

The invention of the laser 25 years ago resulted in powerful light sources which led to the observation of unexpected and striking phenomena. New fields of science such as holography and nonlinear optics developed constituting the basis of this volume. The classical principle of linear superposition of light waves does not hold anymore. Two laser beams crossing in a suitable material may produce a set of new beams with different directions and frequencies.

The interaction of light waves can be understood by considering the optical grating structures which develop in the overlap region. The optical properties of matter become spatially modulated in the interference region of two light waves. Permanent holographic gratings have been produced in this way by photographic processes for many years. In contrast, dynamic or transient gratings disappear after the inducing light source, usually a laser, has been switched off. The grating amplitude is controlled by the light intensity. Dynamic gratings have been induced in a large number of solids, liquids, and gases, and are detected by diffraction, 'forced light scattering' of a third probing beam, or by self-diffraction of the light waves inducing the grating. The combined interference and diffraction effect corresponds to four-wave mixing (FWM) in the language of nonlinear optics. The process is called degenerate if the frequencies of the three incident waves and the scattered wave are equal. Degenerate four-wave mixing (DFWM) is a simple method to achieve phase conjugation, i.e. to generate a wave which propagates time reversed with respect to an incident wave.

Laser-induced dynamic gratings have been studied for the past twenty years but strong interest in the field has occurred only recently due to many scientific and technological applications which have been demonstrated. This volume is intended to collect the most important results in order to achieve a unified treatment of this rapidly expanding field and to indicate directions of further work. Special emphasis is given to the following subjects:

- grating materials and mechanisms of light-induced index and absorption changes: photorefractives, semiconductors, and other systems with large third-order nonlinear optical susceptibilities
- grating and four-wave mixing theory
- investigation of dynamic phenomena by forced light scattering, e.g., relaxation and transport kinetics in solid-state physics, laser-induced ultrasonics, flow studies, and photochemistry
- real-time holography, optical computing, and phase conjugation

– photonic device applications, spatial hole-burning in lasers, distributed feedback, beam deflection, amplification, and modulation.

A sizeable number of researchers now work in the field of dynamic gratings, and we would like to thank them for permission to reproduce their results. We tried to list as many original papers as possible, but the selection of the material is, of course, influenced by the personal view of the authors. We apologize if we did not consider all subjects which might be important to our colleagues.

We finally want to express our sincere thanks to Prof. H. Gerritsen, Brown University, USA, Dr. K. Jarasivnas, Physical Faculty of Vilnivs, and Dr. S. Odulov, Ukrainian Academy of Science, USSR, for very helpful comments on an initial version of the manuscript. Thanks are also due to the IBM publication staff, Rüschlikon, to Mrs. C. Thiel for secretarial support, and to Mrs. A. Rapp of Springer-Verlag. Prof. T. Tamir, the series editor, and Dr. H. Lotsch of Springer-Verlag carefully revised the text and contributed new ideas.

Berlin · Zürich · Rüschlikon   *H. J. Eichler · P. Günter*
October 1985   *D. W. Pohl*

Contents

1. Introduction

In the following chapter the concept of laser-induced dynamic gratings is introduced. It is then described how dynamic gratings are produced and detected experimentally. Grating excitation mechanisms in various materials are surveyed and the importance of dynamic gratings in four-wave mixing and light-scattering experiments is explained. Further applications in holography and phase conjugation are briefly summarized.

1.1 Outline of the Book

The aim of the book is twofold. First, because of the increasing amount of work spent on the subject of laser induced-gratings, it is necessary to collect the most important results to facilitate further work. The parallel discussion of forced light scattering and real time holography should help to exchange ideas between these two fields. Secondly, it is hoped that a unified treatment of the subject starting from simple principles will help to spread the knowledge and to stimulate applications in science and technology.

1.1.1 Permanent and Dynamic Gratings

The optical properties (e.g. refractive index and coefficient of absorption) of matter become spatially modulated in the interference region of two intensive light waves. Permanent gratings have been produced in this way by photographic processes for many years. The experiments of *Wiener* [1.1] giving the first demonstration of standing light waves, the use of this by *Lippmann* [1.2] one year later in the earliest process of color photography, and various holographic experiments following *Gabor* [1.3] provide examples. Besides the well-known silver-halide photographic emulsions, photochromic, thermoplastic and other materials are also used for permanent hologram recording (see for instance, [1.4]).

In contrast, this monograph deals with dynamic or transient gratings which disappear after the inducing light source, usually a laser, has been switched off. These gratings have been produced in a large number of solids, liquids and gases, and are detected by diffraction (forced light scattering) of a probing beam or by self-diffraction of the light waves inducing the grating.

Lasers are almost ideally suited to generate grating-like structures: coherence, collimation and intensity provide strong interference patterns as the source of material excitation, while wavelength tunability and short-pulse capability allow a particular type of light/matter interaction to be selectively addressed.

The diffraction of light from such a grating is closely related to the classical (or spontaneous) scattering of light from random fluctuations. The formation of transient gratings is the basis of real-time holography, phase conjugation, and four-wave mixing.

In presenting the subject, we shall distinguish three processes: pump beam interference, grating formation/material response, and grating detection. The division into these three steps is adequate for most of the work reported here and facilitates the analysis of the experiments described in this book. The three processes are introduced in the next three sections, they are discussed from a phenomenological point of view in Chap. 2, and are then described in more detail in the subsequent chapters.

1.1.2 Grating Creation

In order to create a grating, two beams (usually from the same laser source) are arranged to interfere. For most purposes, it is convenient to use collimated TEM_{00} beams which are close to an ideal plane wave, and provide plane gratings when brought to intersection. Laser-beam propagation and plane-wave interference are briefly reviewed in Sects. 2.1 and 2, respectively.

Interference produces a spatially periodic light intensity and/or polarization distribution which changes the optical properties of a material placed into the interference region. The spatial modulation of the optical material constants acts as a diffraction grating.

1.1.3 Types of Gratings

A laser-induced change of the optical properties is always caused by some material excitation, i.e., deviation from (thermal) equilibrium, which couples to the refractive index and/or absorption coefficient. Absorption of light, for instance, populates excited electronic states in the electron volt (eV) energy range of the sample material in the first place. Thus, inside the interference pattern, a *population density grating* can be created. During the decay of this electronic excitation, various lower-energy electronic, vibrational or other states may become populated forming *secondary gratings*. Of particular importance in this context are *space charge* gratings which build up in photorefractive materials during irradiation. Finally, the excitations thermalize and thus inevitably produce a *temperature grating*. The latter is accompanied by *stress, strain*, and *density* variations. In a mixture of different components, it may even produce a *concentration grating*. All these excitations couple to a refractive index and/or an absorption coefficient, i.e., they all form optical gratings.

A phenomenological discussion of material response is given in Sect. 2.3. The physical mechanisms responsible for grating formation and their dynamics are discussed in detail in Chap. 3.

A wide class of light-induced grating structures is quasi-stationary, i.e., the grating maxima and minima neither oscillate nor propagate. When the exciting radiation is turned off, the amplitude of such gratings decays monotonously, without oscillation. In classical (spontaneous) light scattering, these mechanisms are responsible for the so-called Rayleigh line or central peak, a line in the scattered spectrum centered at the frequency of the incident beam.

Some of the mechanisms responsible for light-induced grating formation, however, involve rapid oscillations of the grating amplitude, mostly connected with wave-like propagation. Density changes, for instance, create sound waves, i.e. acoustical phonons in condensed matter. Similarly, gratings in the molecular vibration amplitudes may be excited, i.e. optical phonons in solids. Such gratings cause the well-known stimulated Brillouin and Raman scattering (Sect. 1.2.2), characterized by a distinct frequency offset between probe and scattered light. A great deal of attention has been paid to Raman and Brillouin processes mostly without explicit use of the grating picture, and a correspondingly complete documentation is available.

The situation is different for laser-induced quasi-stationary gratings. The present monograph focuses on such quasi-stationary gratings in a first attempt to provide a unified and – as far as possible – complete coverage of the subject; oscillatory gratings will be discussed only to the extent necessary to exhibit common features and to give some typical examples (Sects. 4.7, 5.3).

1.1.4 Grating Detection

Gratings are generally detected by diffraction of a probing light beam. The different methods employed are outlined in Sect. 2.4. Diffraction theory, as applicable to laser-induced gratings, is reviewed in Chap. 4. The theory starts in a way similar to permanent, e.g. holographic gratings, but must also include the effect of self-diffraction, where it is not only necessary to describe the action of the grating on a light wave, but also the simultaneous production of the grating by the light wave resulting in a nonlinear problem of wave propagation. This situation is analogous to the propagation of coupled waves in nonlinear optical materials, briefly reviewed in Sect. 2.7.

The diffracted light can either be recorded directly or heterodyned with a split-off beam from the probe light. The first technique has the advantage of simplicity, but requires sufficiently strong signals. The second one offers increased sensitivity, but requires a larger experimental effort. The different detection techniques will be discussed in Sects. 2.4.4 and 5.

1.1.5 Four-Wave Mixing

The nonlinear interaction of four light-waves has been studied in a large number of experiments. An arrangement is of special interest, where two antiparallel pump beams are mixed with a third optical (object) beam having another propagation direction in a dynamic recording medium. All beams have the same frequency (degenerate four-wave mixing, DFWM). In the simplest approach, one of the antiparallel beams forms a grating with the third beam which is read-out by the second pump beam having reverse propagation direction to the first one.

The diffracted (fourth) wave propagates antiparallel to the object beam with very unusual image-transformation properties. The new beam exactly retraces the path of the object wave, and any phase aberrations which occurred to the object beam are cancelled out in the diffracted wave as it again passes the aberration source (optical phase conjugation). Therefore, by using four-wave mixing in a dynamic medium, high quality optical beams can be double-passed through a poor quality or temporarily fluctuating optical system with no loss in beam quality.

This example shows that diffraction at laser-induced gratings is analogous to the nonlinear mixing of light waves, as can be worked out further by using the concept of moving gratings.

Four-wave theory will be discussed in context with the general theory of diffraction in Chap. 4. Real-time holography and phase-conjugation, the applied side of four-wave mixing, will be discussed in detail in Chap. 6. The historical development of these fields will be described in Sect. 1.2.5.

1.1.6 Forced Light-Scattering

Besides being an interesting physical phenomenon on its own, diffraction at laser-induced gratings led to the observation of new phenomena and allowed the detection and study of other effects in greater detail and more conveniently than possible with conventional methods. The progress is evident by comparison with the established classical light scattering techniques (Sect. 2.6).

Spontaneous light scattering (e.g., Rayleigh scattering) takes place at weak and random fluctuations of macroscopic material properties (e.g., entropy or temperature). The transient grating technique replaces such fluctuations by strong and coherent excitations from a laser-induced grating. The properties of such an excitation can be investigated by the forced scattering of a probing beam, thus obtaining signals much larger than in spontaneous scattering experiments.

Diffraction at laser-induced gratings has therefore developed into a method for investigation of optically excited materials. The detection of second sound in NaF, the measurement of anisotropic heat conduction in liquid crystals (Sect. 3.7.1), the measurement of surface recombination velocities in semiconductors (Sect. 5.5) and the detection of exciton diffusion in organic and inorganic crystals (Sect. 5.6) may serve as first examples.

The fundamental reasons for the usefulness and high sensitivity of forced scattering are:

1) Spatially modulated excitations can be detected optically with larger sensitivity than spatially homogeneous ones.

2) Transport phenomena, such as diffusion, need modulated excitation to become detectable at all.

3) Forced fluctuations have amplitudes larger by many orders of magnitude than thermodynamic fluctuations.

4) Grating-like sinusoidal modulation of the refractive index lends itself particularly well to interpretation because the interaction with a probing light beam is well defined and easy to evaluate.

5) The decay of a grating after pulsed excitation and the corresponding time constant are frequently the most important information to be obtained from a transient grating experiment. Depending on the nature of the grating, the decay time may vary between picoseconds (and less) and seconds (and more). Experimental requirements therefore vary considerably for different applications. Forced scattering further allows one to distinguish between different decay mechanisms. For example, if the decay takes place by local relaxation, the observed time constant is independent of the grating constant q while linear diffusive transport yields a quadratic dependence on q.

6) Two other important parameters obtainable from grating experiments are the coupling constants between the material excitation and the pump light on the one hand, and the probe light on the other. The two constants and their variations (say, with temperature, pressure, or sample orientation), relate to the interaction strength of a material excitation and electromagnetic radiation.

1.1.7 Applications in Holography and Phase Conjugation

On the technical side, laser-induced transient gratings can be used in holography for real-time processing of optical fields (Chap. 6). Real-time holography was suggested already in 1967. In recent years, this subject has regained interest because of the demonstration of phase conjugation or time reversal of optical wave fronts [1.5]. Using this process, it is possible to design self-adaptive optical systems which compensate time-varying phase distortions in high-gain laser oscillators and amplifiers, or in optical transmission lines through atmosphere, water, or optical fibers. At present, real-time holography is a rapidly growing field of research.

In addition to phase conjugation, infrared-to-visible image conversion, image amplification and optical logic (addition, subtraction, convolution, correlation) have been demonstrated. Strong interactions between pump and probe beams in suitable materials are required for efficient operation of devices performing such optical functions.

1.2 Historical Development

Historically, dynamic optical gratings were first produced not by lasers or other light sources but by ultrasonic waves. Early examples of such transient gratings investigated by optical probes are given in the ultrasound-induced light diffraction experiments of *Debye* and *Sears* [1.6a], and *Lucas* and *Biquard* [1.6b]; velocity, dispersion, and damping of ultrasonic waves could be determined this way. Ultrasound gratings are now widely used for light beam deflection, scanning and modulation.

Interfering light waves are just as well suited to create grating-like structures in a medium, but conventional light sources are too weak and too incoherent to produce sizeable amplitudes except in very strong interactions such as in the photographic process. The situation changed when lasers became available.

The invention of the ruby laser in 1960, and a large variety of other pulsed high-power and cw lasers in the following years led to the observation of unexpected, striking phenomena and the development of new fields of optical science. High-power pulsed lasers produced optical harmonics, stimulated scattering processes and other new nonlinear effects. The extraordinary coherence of lasers compared to classical light sources made possible the development of holography from a scientific curiosity to a standard technique. On the basis of laser light sources, various independent paths led to light-induced gratings and to their theoretical understanding. Light-induced grating phenomena became of interest in various laser-related fields so that the following historical sketch may be somewhat incomplete and influenced by the personal view of the authors.

1.2.1 Hole Burning in Lasers

Light-induced gratings are produced by the standing light wave in a laser cavity. The light wave burns spatial holes into the inversion sustaining laser action. The concept of spatial hole-burning was introduced by *Haken* and *Sauermann* [1.7] and *Tang* and *Statz* [1.8] to describe the statistical temporal spikes observed in the output power of pulsed solid-state lasers. Later, spatial hole-burning was recognized as a limiting factor in achieving high power, single-frequency emission from cw dye lasers. The problem was solved using ring lasers with traveling-wave operation [1.9]. The first direct experimental proof of spatial hole-burning was given by *Boersch* and *Eichler* [1.10] in 1967 who detected the corresponding optical grating by diffraction of a second laser (Figs. 1.1, 2).

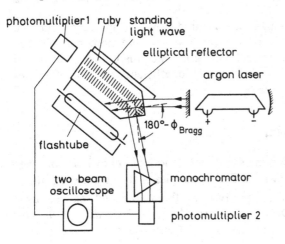

Fig. 1.1. Experimental setup to investigate diffraction at a grating formed by spatial hole-burning in a ruby laser. The figure is not true to scale. The ruby rod was 10 cm long and 1 cm in diameter

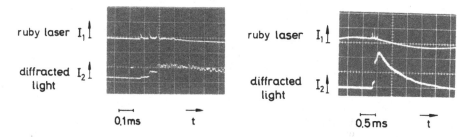

Fig. 1.2. Diffraction at a laser-induced grating produced in the setup shown in Fig. 1.1. Ruby laser output I_1 (3–4 spikes) and diffracted light I_2 with different time resolution. The I_2 trace is disturbed by some slow background

1.2.2 Stimulated Scattering

A particularly striking phenomenon in the propagation of laser beams is the apparent high reflectivity of normally transparent materials when hit by a sufficiently powerful pulse. The reflected light goes straight back in the direction opposite to the incident beam. Moreover, the reflected pulse usually has a frequency slightly offset from that of the incident radiation.

After the discovery of these phenomena, it soon evolved that the reflectivity was caused by strong transient gratings. These gratings build up from random fluctuations by the interaction of the incident and the classically scattered light. Depending on the dominant type of interaction, gratings of optical phonons, gigahertz acoustic waves (hyper-sound), or temperature (entropy) are created. In analogy to classical light scattering, these processes were called stimulated Raman scattering (SRS), stimulated Brillouin scattering (SBS), and stimulated thermal Rayleigh scattering (STS or STRS), stimulated Rayleigh wing scattering (SRWS) and stimulated concentration scattering (SCS). The abbreviations and their meanings are summarized in Table 1.1.

Table 1.1. Common abbreviations for stimulated scattering processes

Abbrev.	stands for...	originates from fluctuations in...
STS	stimulated thermal scattering	entropy, temperature
STRS	stimulated (thermal) Rayleigh scattering	entropy, temperature
SCS	stimulated concentration scattering	concentration
SRWS	stimulated Rayleigh wing scattering	orientation of molecules
SBS	stimulated Brillouin scattering	sound waves, acoustical phonons
SRS	stimulated Raman scattering	molecular vibrations, optical phonons

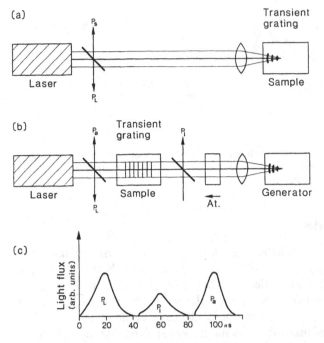

Fig. 1.3a–c. Stimulated scattering. (**a**) oscillator arrangement; scattered beams built up from noise. (**b**) amplifier arrangement: superposition of two well-defined beams in the sample cell. AT = Attenuator (**c**) laser pulse, incident and amplified scattered pulses measured with arrangement (**b**)

A typical experimental arrangement for stimulated scattering is sketched in Fig. 1.3a. A powerful laser pulse is focused on the sample material. Spontaneously scattered light is first emitted in all directions and thus interferes with the incident pulse producing a manifold of superposed grating structures. The latter couple to material excitations with appropriate properties. For example, a (travelling) grating pattern originating from Brillouin scattering will enhance the sound wave with the same wave-vector and velocity. The enhanced material excitation in turn tends to enhance the scattered intensity in the corresponding direction.

The strongest enhancement occurs for the longest scattered light paths within the scattering volume. In the simple arrangement of Fig. 1.3a, the optimum enhancement paths point in the forward and backward direction. Scattered radiation travelling parallel or antiparallel to the incident laser pulse can therefore build up an intensity comparable to that of the incident pulse if a certain intensity threshold is exceeded. The intensity of the incident pulse gets reduced correspondingly. For this reason, scattering into less favorable directions actually gets suppressed. Eventually, in this competitive process called stimulated scattering, either the forward or the backward direction survives. Which of the two dominates depends, among others, on the pulse duration t_p.

If the physical length of the pulse, ct_p/n (c/n: velocity of light inside sample) is (considerably) smaller than the sample length, stimulated forward scattering prevails because the scattered pulse overlaps with the incident one during the whole time of flight through the sample (if dispersion is ignored). If, on the other hand, ct_p/n is larger than the sample thickness, the backward travelling scattered pulse also overlaps with the incident one on the whole path. The backward pulse has the further advantage that it interacts with just arriving, undepleted incident radiation. Therefore, it may become a very intense compressed pulse travelling right back to the laser source. This situation is sketched in Fig. 1.3a. The beam splitter BS separates part of the incident (P_L) and part of the backward-scattered (P_S) from the main light path for detection and analysis.

Although they have initiated the field of laser-beam-induced gratings, these so-called generator experiments have been used seldom for grating investigations. We shall not go into the details of stimulated scattering, but refer the reader to review the available articles [1.11–15].

Further development moved in two directions:

1) SRS with picosecond pulses.
2) Replacement of spontaneously scattered light as initiator by a well-defined second pump light source.

In Technique 1, the transient grating moves along the sample together with the collinearly propagating incident and scattered (Raman) pulses. Since an almost complete conversion into Raman light is achieved in this way, this technique is used for frequency shifting of laser pulses. Properties of the associated transient gratings have been of minor interest in this context.

Technique 2 was introduced by *Bloembergen* et al. [1.16] for SRS, by *Denariez* and *Bret* [1.17] for SBS and SRWS, and by *Pohl* et al. [1.18] for SBS and STS. In a combination of stimulated scattering and amplification, the reflected beams from SRS, SBS, STS, or simply from a mirror were attenuated and brought to interaction with the incident beam in a second cell. Interference of the two beams inside the sample creates a well-defined transient grating which manifests itself in an amplification of the reflected beam at the expense of the incident one. Fig. 1.3b shows such an amplifier experiment. The attenuator At is unidirectional in order to adjust the intensity of P_i to a convenient level.

Figure 1.3c depicts the pump (P_L), incident scattered (P_i) and amplified scattered (P_a) pulses measured with the arrangement of Fig. 1.3b [1.18]; the temporal separation was achieved by using optical delay lines. The amplification by a factor of approximately 2 is clearly seen.

The results from an amplifier experiment were much easier to interpret than the previous stimulated scattering data. However, all the information was contained in an intensity difference of a light pulse in front of and behind the amplifier cell. Differences, however, are always somewhat difficult to measure reliably, in particular if they become small.

1.2.3 Forced Scattering or the Transient Grating Technique

The next logical step was to separate grating generation and detection. A third probe beam was required for this purpose, thus establishing the forced Rayleigh scattering (FRS) or transient grating technique. The latter was first used by *Eichler* and co-workers in 1972 [1.19] to determine heat diffusivities in ruby and glycerol. Independently, *Pohl* and co-workers [1.20] developed FRS as an extremely sensitive tool to study the propagation of heat in NaF at low temperatures, in particular, the transition to second sound.

This low or thermal energy FRS was subsequently applied to problems of heat and mass diffusion and low-energy excitations in amorphous materials. Liquid crystals, polymers, critical mixtures, and glasses at low temperatures attracted particular attention because the information obtained by FRS was not easily accessible by other means. *Woerdman* [1.21] first used the transient grating technique for the investigation of electronic (high-energy FRS) processes in semiconductors [1.15]. FRS at electronic excitations was combined by *Phillion* et al. [1.22] with the picosecond technique to cope with the short lifetimes involved.

1.2.4 Light-by-Light Scattering

Another source for the development of the forced light-scattering technique is the idea of light-by-light scattering. In classical linear optics, the intensities of light waves are small so that interaction can be neglected, i.e., the principle of undisturbed superposition can be applied. When high-power lasers became available, it was observed that the optical properties of matter are changed by a laser beam, so that light-light interaction becomes important. The first demonstration of such an interaction, which can be interpreted as self-diffraction at a laser-induced grating, was given in 1966 by *Chiao, Kelley* [1.23a], and *Garmire* and *Carman* [1.23b]. The work of *Boersch* and *Eichler* [1.9], which has already been mentioned in connection with spatial hole-burning, was also aimed at showing diffraction of light at a standing light wave; this work has used the grating description for the first time in this connection.

1.2.5 Holography, Phase Conjugation, and Four-Wave Mixing

Holography is another important field which has stimulated interest in laser-induced grating phenomena. A grating can be considered as an elementary hologram, which is similarly produced by interference of a reference and an object wave as a hologram in general. Holograms are usually stored permanently in photographic or other suitable materials, but in 1967 *Gerritsen* et al. [1.24] suggested holograms, which are stored only temporarily, for real-time processing of changing optical fields. Experimentally, a dye solution was used to record temporary holograms. Imaging and magnification were demonstrated to show the feasibility of a holographic microscope with large depth of field.

Further early work on real-time holography was done by *Woerdman* [1.21] with Nd:YAG lasers and silicon as recording material. This work stimulated considerable interest in semiconductor gratings.

The number of papers dealing with real-time holography increased considerably in recent years, after *Stepanov* et al. [1.25] and *Zel'dovich* et al. [1.26] suggested using this process for cancellation of laser beam distortions due to propagation through phase-disturbing media [1.5]. This could be of relevance for atmospheric or complex optical communication channels. Independently, *Yariv* [1.27] and *Hellwarth* [1.28] proposed a nonlinear optical method to cancel beam distortion. The method was named phase-conjugation, and was demonstrated experimentally utilizing various nonlinear optical processes, e.g., three-wave mixing, four-wave mixing, stimulated Brillouin scattering and photon-echoes.

In photorefractive materials, large photo-induced refractive index changes can be induced even with low power cw laser beams. This allows cancellation of beam distortions with simultaneous amplification [1.29]. Figure 1.4 shows an early phase conjugation experiment using $KNbO_3$ as the photorefractive material [1.20]. A diffraction limited optical beam (A) is phase disturbed by a distorting glass plate (B), but the phase conjugated signal (C) regains the original beam quality, as it passes again through the distorting plate (D). It was possible to use a low-power cw laser beam in this case because of the strong photorefractive interaction occurring in $KNbO_3$.

It should be emphasized that phase conjugation by four-wave-mixing or by real-time holography are just different names for the same phenomenon. The term real-time holography is more inclusive being used to describe the whole class of pictorial information-processing techniques based on instantaneous or

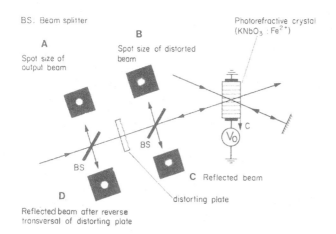

Fig. 1.4. Real-time correction of phase distortions by four-wave mixing in dynamic recording media, e.g., photorefractive crystals

near-instantaneous optical response. These techniques almost always involve a grating-like modification of the optical properties (the complex index of refraction) of the hologram medium by two incident optical beams and a simultaneous or slightly delayed scattering of a third wave at the hologram grating. Reviews of phase conjugation and real-time holography have been given in [1.5, 27, 31–33].

2. Production and Detection of Dynamic Gratings

Dynamic gratings are produced by interfering light beams usually derived from lasers. We therefore recapitulate some basic properties of laser beams and discuss two-beam interference more extensively. The next step to be considered in grating production, is the material response which is treated here only phenomenologically and will be discussed in more detail in Sect. 3.3. Gratings are detected by diffraction which is described theoretically in Sect. 4.3. Some important results and experimental considerations, however, are already given here.

In addition, the relevance of dynamic gratings in light scattering experiments and the general context of nonlinear optics are discussed in Sect. 2.5.

2.1 Laser Beams and Pulses

Since all the grating physics in this monograph is based on laser excitation and detection, it is useful to review the parameters of laser beams and their propagation [2.1]. Here, it is sufficient to consider the fundamental TEM_{00} mode which comes close to the ideal plane wave

$$E(r, t) = \frac{A}{2} \exp [i(k \cdot r - \omega t)] + c.c. \quad . \tag{2.1}$$

E is the instantaneous and local electric field vector, A is the wave amplitude, k its wave vector, and $\omega = 2\pi\nu$ its angular frequency. r and t have the usual meaning of spatial coordinate vector and time, respectively, while c.c. stands for the conjugate complex. The TEM_{00} mode has a Gaussian rotationally symmetric amplitude distribution

$$A(\varrho) = A_0 \exp(-\varrho^2/w^2) , \tag{2.2}$$

where ϱ is the cylindrical coordinate perpendicular to the direction of propagation z' and w is called the spot size (Fig. 2.1).

In SI units, the intensity I of a light beam at point r, time t is

$$I(r, t) = \frac{1}{2} \varepsilon_0 c n |A(r, t)|^2 = \frac{1}{2Z} |A|^2 , \tag{2.3}$$

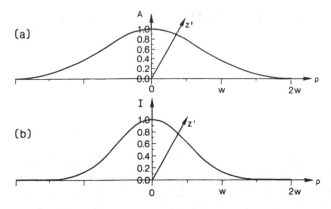

Fig. 2.1. (a) Electric field distribution, (b) intensity distribution of TEM$_{00}$ mode

where c is the velocity of light, ε_0 the vacuum permittivity, n the refractive index and Z the wave resistance of the material. The intensity distribution of a Gaussian beam is (Fig. 2.1b)

$$I(\varrho) = I_0 \exp\left(-2\varrho^2/w^2\right) . \tag{2.4}$$

At $\varrho = w$, the electric field has reached $1/e \simeq 37\%$ of its maximum value A_0 while the intensity is at $1/e^2 \simeq 14\%$ of I_0 already. The total power or light flux P_t of a TEM$_{00}$ beam is

$$P_t = 2\pi \int_0^\infty I(\varrho)\varrho\, d\varrho = (1/2)\pi w^2 I_0 . \tag{2.5}$$

Roughly 90% of this flux is contained within a radius equal to the spot size w.

The laser beam diameter changes during propagation. Thus, except when going through a focus, the wave fronts are not perfectly plane and the spot size is not constant but a function of z'. Since the divergence is inversely proportional to the beam diameter, a sufficiently large spot size $w \gg \lambda$ is a necessary requirement for plane-wave behavior.

Short laser pulses are frequently used for grating excitation and detection. If their duration t_p is sufficiently small, the total pulse energy per unit area E_p, i.e. the exposure or fluence

$$E_p = \int_{-\infty}^{+\infty} I\, dt \tag{2.6}$$

and the total laser pulse energy

$$W = 2\pi \int_0^\infty E_p \varrho\, d\varrho = \int_{-\infty}^{+\infty} P_t\, dt \tag{2.7}$$

are more relevant parameters than instantaneous intensity and flux.

2.2 Two-Beam Interference and Interference Gratings

Two-beam interference produces a spatially modulated light field which is called an interference grating. In many textbooks [2.5] interference of two plane light waves with parallel polarization is considered. Excitation of dynamic gratings, however, is also possible by interfering beams with different, e.g. perpendicular polarization. Therefore, the general case of superposition of beams with different polarization is also treated here. This leads to an interference grating with an amplitude described by an interference tensor. The tensorial description of two-beam interference is closely related to the Taylor expansion of the nonlinear optical polarization in terms of products of the electrical field components.

2.2.1 Superposition of Two-Plane Waves

The experimental arrangement for the production of laser-induced gratings is conceptionally simple although technically it is sometimes quite demanding. The principal setup is sketched in Fig. 2.2. Light from a more-or-less powerful pump laser is split into two-beams A and B with wave vectors k_A and k_B, electric field amplitudes A_A, A_B, and intensities I_A, I_B. The two-beams intersect at an angle θ at the sample and create an interference pattern, the grating vector q of which is

$$q = \pm (k_A - k_B) \ . \tag{2.8}$$

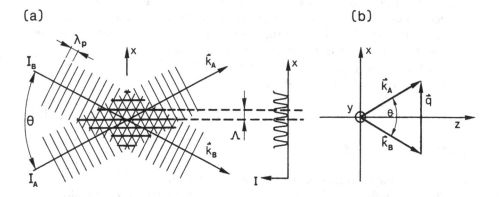

Fig. 2.2a,b. Grating production by interference of two light waves with intensities I_A, I_B and wave vectors k_A and k_B

The spatial period Λ is

$$\Lambda = 2\pi/q , \tag{2.9}$$

where $q = |\boldsymbol{q}|$. Λ can be expressed in terms of the pump wavelength λ_p and the angle θ

$$\Lambda = \lambda_p/2 \sin(\theta/2) . \tag{2.10}$$

For small angles θ, the grating period is approximately

$$\Lambda \approx \lambda_p/\theta \quad \text{for} \quad \theta \ll 1 . \tag{2.11}$$

Note that, up to now, the wave vectors $\boldsymbol{k}_a$, $\boldsymbol{k}_b$, wavelength λ_p and intersection angle θ are measured in the material with refractive index n. For nearly normal incidence, (2.11) is also approximately valid if the wavelength $\lambda_{0p} = n\lambda_p$ and the intersection angle $\sin\theta_0 = n\sin\theta$ are measured outside the sample

$$\Lambda \approx \lambda_{0p}/\theta_0 \quad \text{for} \quad \theta_0 \ll 1 . \tag{2.12}$$

By varying the intersection angle θ_0 the grating period Λ can be changed. The maximum value of Λ is limited by the diameter of the laser beam inducing the grating. Experimentally, values up to approximately 100 µm have been used. The smallest grating-period values are achieved when the two excitation beams are antiparallel with $\theta = 180°$ giving a minimum value of $\Lambda = \lambda_p/2 = \lambda_{0p}/2n$. Using a visible laser and highly refractive material, the grating period may be as small as 100 nm.

Cartesian coordinates will be chosen throughout this volume such that the x- and z-axes are in the plane defined by $\boldsymbol{k}_a$, $\boldsymbol{k}_b$. For symmetry reasons, these vectors point into the directions of the two bisectrix between the pump beams as indicated in Fig. 2.2b. The y-direction points upward so as to produce a right-handed system. Thus the x-axis is parallel to $\boldsymbol{q}$ while the z-axis, for small θ_p, almost coincides with $\boldsymbol{k}_A$, $\boldsymbol{k}_B$, the propagation directions of the pump beams. The cross-section of the interference pattern is confined to the xy plane being perpendicular to $(\boldsymbol{k}_A, \boldsymbol{k}_B)$ and containing $\boldsymbol{q}$.

$\boldsymbol{k}_a$, $\boldsymbol{k}_b$, and $\boldsymbol{q}$ can be expressed as

$$\boldsymbol{k}_{A,B} = \boldsymbol{z}_0 k_z \pm \boldsymbol{x}_0 k_x , \tag{2.13}$$

$$\boldsymbol{q} = \pm \boldsymbol{x}_0 q = \pm \boldsymbol{x}_0 2k_x , \tag{2.14}$$

where $\boldsymbol{x}_0$, $\boldsymbol{z}_0$ (and $\boldsymbol{y}_0$) denote the respective unit vectors.

The electric-field amplitude distribution inside the interference region is

$$A = A_A e^{+ik_x x} + A_B e^{-ik_x x} , \tag{2.15}$$

and the total time-dependent field $E(r, t)$ is given by

$$E(r, t) = \frac{A}{2} e^{i(k_z z - \omega_p t)} + \text{c.c.} \ . \tag{2.16}$$

The intensity distribution is

$$I = \frac{n}{2} \varepsilon_0 c (A \cdot A^*) = \frac{n}{2} \varepsilon_0 c (|A_A|^2 + 2 A_A \cdot A_B^* \cos 2 k_x x + |A_B|^2)$$

$$= I_A + 2 \Delta I \cos 2 k_x x + I_B \ , \quad \text{where} \tag{2.17}$$

$$\Delta I = \frac{n}{2} \varepsilon_0 c A_A \cdot A_B^* \tag{2.18}$$

is the intensity modulation amplitude and the asterisk denotes the complex conjugate (c.c.).

ΔI is the significant parameter for optical grating creation if both sample and interaction mechanism are isotropic. In anisotropic media, however, or with anisotropic interaction, gratings may also be induced if $A_A \perp A_B$ and $\Delta I = 0$. To account for such a situation, we introduce the *interference tensor* ΔM which is defined for vacuum as

$$\Delta M_{ij} = \tfrac{1}{2} \varepsilon_0 c A_{A,i} A_{B,j}^* \ . \tag{2.19}$$

ΔI is the absolute value of the trace of ΔM, i.e.,

$$\Delta I = |\text{tr}\{\Delta M\}| \ . \tag{2.20}$$

2.2.2 Superposition of Short Pulses

Present-day mode-locked lasers provide pulses of very short duration, say 1 ps, equivalent to a physical length of 0.3 mm. The interference of two beams derived from such a source depends on the delay between the pulse fractions travelling along paths A and B. The time dependence of such pulses is close to Gaussian with half width t_p

$$I_{A,B}(t) = \hat{I}_{A,B} \exp\left\{ -[(t \pm \tau/2)/t_p]^2 \right\} \ , \tag{2.21}$$

where τ is the delay of pulse B with regard to A, and $\hat{I}_{A,B} = \tfrac{1}{2} n c \varepsilon_0 \hat{A}_{A,B}^2 / 2$ are the peak powers of the two pulses. The magnitude of the interference tensor in this case also depends on the overlap of the two pulses given by the ratio τ/t_p, i.e.,

$$\Delta M_{ij} = \tfrac{1}{2} c \varepsilon_0 \hat{A}_{A,i} \hat{A}_{B,j}^* \exp\left[-(\tau/2 t_p)^2 \right] \exp\left[-(t/t_p)^2 \right] \ . \tag{2.22}$$

Thus the temporal behavior of $\Delta M(t)$ is the same as that of the original pulse(s), but its amplitude decreases in proportion to $\exp\left[-(\tau/2 t_p)^2 \right]$.

2.2.3 Superposition of Beams with Different Polarizations

We discuss four important special cases here:

1) s polarization: $A_A \| A_B \| y_0$ (Fig. 2.3a). This is probably the most frequently used experimental situation and also the simplest one. ΔM degenerates into the one-element tensor

$$\Delta M = \begin{pmatrix} 0 & 0 & 0 \\ 0 & \Delta I & 0 \\ 0 & 0 & 0 \end{pmatrix}, \tag{2.23}$$

where, in this case $\Delta I = \sqrt{I_A I_B}$.

If, in addition, $I_A = I_B$ then $\Delta I = I_A$ and

$$I = 2 I_A (1 + \cos qx) . \tag{2.24}$$

Thus, the intensity is fully modulated, varying between zero and four times the value for a single beam. This case is favorable for most of the applications to be discussed.

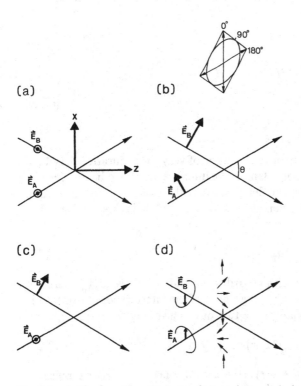

Fig. 2.3a–d. The four arrangements of pump beam polarization discussed in the text

2) p polarization: A_A, $A_B \perp y_0$ (Fig. 2.3b). In this case, A_A and A_B are in the xz-plane with A_{Ax}, A_{Bx}, $A_{Bz} \geq 0$ but $A_{Az} < 0$. The interference tensor becomes

$$\Delta M = \tfrac{1}{2} c\varepsilon_0 \begin{pmatrix} A_{Ax}A_{Bx}^* & 0 & A_{Az}A_{Bx}^* \\ 0 & 0 & 0 \\ A_{Ax}A_{Bz}^* & 0 & A_{Az}A_{Bz}^* \end{pmatrix} \tag{2.25}$$

corresponding to an intensity modulation

$$\Delta I = |\mathrm{tr}\{\Delta M\}| = \tfrac{1}{2} c\varepsilon_0 |A_{Ax}A_{Bx}^* + A_{Az}A_{Bz}^*| . \tag{2.26}$$

Note that the second term in (2.26) is negative. The physical interpretation of ΔM is as follows: Depending on the relative phase $\Phi_{AB} = qx$ of the two pump beams along x, the superposition of A_A and A_B results in a polarization varying between linear (for 0, 180° relative phase) and elliptic, as sketched in the inset to Fig. 2.3b. It is seen that the intensity modulation ΔI disappears completely at $\theta = 90°$ since $A_A \perp A_B$ in this case. The interference field polarization is particularly interesting if, in addition, $|A_A| = |A_B|$. It then points into the x-direction for $\Phi_{AB} = 0$, becomes circular at $\Phi_{AB} = \pi/2$, and finally, at $\Phi_{AB} = \pi$, it is linear in the z-direction, i.e., a longitudinal field with regard to the interference pattern. Interestingly, the polarization interference pattern can be made visible by placing a dichroic medium such as a polaroid foil into the zone of interaction. Thus, for the investigation of optically anisotropic media, perpendicular polarization can be of interest.

3) Mixed polarization: $A_A \parallel y_0$, $A_B \perp y_0$ (Fig. 2.3c). In this case, the electric fields of the excitation beams are perpendicular ($A_A \perp A_B$) for any value of θ. The interference tensor is

$$\Delta M = \tfrac{1}{2} c\varepsilon_0 \begin{pmatrix} 0 & 0 & 0 \\ A_A A_{Bx}^* & 0 & A_A A_{Bz}^* \\ 0 & 0 & 0 \end{pmatrix} . \tag{2.27}$$

No intensity modulation exists. The polarization undergoes periodic changes similar to the previous case. The electric vector is confined to the (y_0, A_B) plane.

4) Opposite circular polarizations and $|A_a| = |A_b|$; $A_{A,B} = (x_0 \cos\theta/2 \pm iy_0 \mp z_0 \sin\theta/2)|A_A|/\sqrt{2}$ (Fig. 2.3d). The interference tensor is

$$\Delta M = \tfrac{1}{4} c\varepsilon_0 |A_A| \begin{pmatrix} \cos^2\dfrac{\theta}{2} & i\cos\dfrac{\theta}{2} & \dfrac{1}{2}\sin\theta \\[2mm] i\cos\dfrac{\theta}{2} & -1 & i\sin\dfrac{\theta}{2} \\[2mm] -\dfrac{1}{2}\sin\theta & -i\sin\dfrac{\theta}{2} & -\sin^2\dfrac{\theta}{2} \end{pmatrix} . \tag{2.28}$$

The intensity modulation is

$$\Delta I = |\mathrm{tr}\{\Delta M\}| = \tfrac{1}{2} c\varepsilon_0 |A_A|^2 \sin^2 \theta \; , \tag{2.29}$$

which becomes vanishingly small for $\theta \to 0$, while the polarization tends to become linearly polarized and rotating with the grating period Λ across the grating structure, i.e., along the x-axis as indicated in Fig. 2.3d. This choice of polarization is obviously favorable for studying optically active media or interactions.

2.2.4 Finite Size Effects

The finite cross-section of the pump-laser beams limits the lateral extent of the interference zone. Hence, the electric-field amplitudes and intensities in (2.14–29) are slowly varying functions of x, y, z in addition to the modulation in the x-direction. Calculation of the spatial variation is straightforward assuming

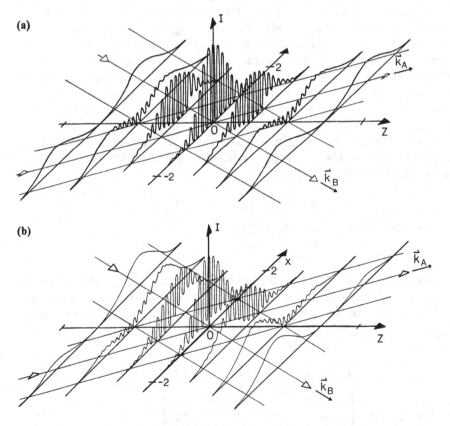

Fig. 2.4a, b. Sketch of the intensity distribution within the volume-grating interference pattern formed by two intersecting Gaussian beams (a) equal intensity, (b) $I_A = I_B/4$ (reproduced, with minor changes of notation, from [2.2])

TEM_{00} beams, but involves a lengthy notation. A quantitative description is given by *Brayton* [2.2] and *Siegman* [2.3]. The interference zone is depicted in Fig. 2.4 (reproduced from [2.2]) for equal and unequal intensities, respectively.

The interference between two TEM_{00} beams will obviously come close to an ideal plane grating if the following three conditions can be met:

1) The minimum width of the interaction zone must be large compared to the grating period, i.e.,

$$qw \gg 1 \ . \tag{2.30}$$

2) The overlap length z_0 of the two beams in z-direction must be large compared to the sample thickness d,

$$z_0/d \gg 1 \ . \tag{2.31}$$

3) The attenuation of the exciting beams must be negligible within the sample, i.e.,

$$Kd \ll 1 \ , \tag{2.32}$$

where K is the absorption constant of the sample material at wavelength λ_p. The first condition puts a limit on focusing of the pump beams to increase intensity; the second puts a limit on the angle θ_p between the beams. The third puts a limit on pump beam utilization by absorption.

The description of a laser-induced grating is particularly straightforward if the above three conditions are satisfied experimentally. In the following discussion, we shall assume this to be the case unless mentioned otherwise. Note that the results stay qualitatively correct even if one or several of the conditions are satisfied marginally only.

Extremely large values of Λ would require inconveniently small angular beam separation θ_p. Under such circumstances, it is possible to use one pump beam only, and to produce the grating by insertion of a comb-like aperture. Values of Λ up to 4 mm were obtained in this way (Sect. 5.2). There is no obvious upper limit to the period of gratings produced in this manner – except that the laser beam cross-section has to be increased proportionally, thus lowering the intensity available for the pumping process.

The grating is stationary in position when the two beams have the same frequency. A frequency offset between the two excitation beams provides traveling grating structures, to be discussed in Sect. 4.6.

2.3 Material Response: Amplitude and Phase Gratings

The mechanisms of light-induced changes of optical materials properties are often described by two steps. First the light produces some material excitation which then leads to a change of the optical properties. In the simplest case the absorption and the refraction of the material are changed resulting in amplitude and phase gratings.

2.3.1 Material Excitation Gratings

When a material is placed into the interference region of the pump waves, some light-matter interaction such as absorption creates a corresponding spatial modulation (grating) of some material property, e.g., the population of an excited electronic state, the conduction electron density (in a semiconductor), the space charges and their accompanying fields (in photorefractive materials), or the temperature, the molecular orientation (fluids) and the concentration (in mixtures).

Many of these changes can be described by the population of one, several, or a whole continuum of excited (e.g., electronic or phonon) states of the sample material. Hence, the corresponding gratings can also be considered population gratings in a generalized sense.

The description in terms of excited-state populations is necessary if the local population density is out of thermal equilibrium. This is usually the case if the excited-state energy is far above the thermal energy $k_B T$ which at room temperature is ca. 25 meV. Strong deviations from thermal distribution can also occur during radiationless decay from the primarily excited electronic state. In solids the energy freed during such a process may create hot phonons which, in turn, decay into cooler ones until thermal energies are reached. This process is very fast because hot-phonon lifetimes are on the sub-picosecond scale. Since today's mode-locked lasers provide pulses down to about 30 femtosecond duration [2.4], such transient effects can play a role in experiments with extremely high time resolution. In other materials, it is also possible that long-lived intermediate states of different nature (e.g., two-level states in glasses, Sect. 5.2) get populated during the decay, particularly at low temperatures. This can considerably slow down the thermalization process, giving rise to secondary grating structures with their own characteristic properties and decay times.

Once the absorbed energy is thermalized locally, the description of the resulting grating in terms of the usual thermodynamic variables, temperature, concentration etc., is appropriate and convenient. The sample as a whole is not in equilibrium as long as these quantities still vary spatially. Their equilibration requires transport of heat, matter, etc. which usually occurs by diffusion. Thus, their decay time depends on the size of the excitation gradients and hence the q vector of the grating.

Note that a diffusion process, in general, does not change the center position of the excited region but tends to smear out its spatial profile. Hence, a grating stays stationary during diffusive decay, i.e., its phase stays constant while its amplitude decreases monotonically.

The dependence of the material excitation on the light intensity depends on its dynamics and cannot generally be expressed by a simple function. The time dependence of the material excitation is often preferably described by some differential equation with the pump light intensity as a source term, for instance, in the homogeneous heat-transport equation.

Under stationary conditions, the material excitation amplitude ΔX is proportional to the modulated intensity amplitude ΔI in the simplest case

$$\Delta X = g^{\mathrm{p}}(\lambda_{\mathrm{p}}) \Delta I(\lambda_{\mathrm{p}}) \ , \tag{2.33a}$$

where g^{p} is a coupling coefficient which depends on the type of material excitation and the pump wavelength λ_{p}. The right-hand side may be considered as the first term of a power series describing the general relations between ΔX and ΔI.

Depending on the nature of excitation, ΔX can be a scalar (temperature, etc.), vector (electric field, flow velocity) or tensor (stress, strain, orientational distribution of excited molecules). Thus, it is convenient for further discussions to rewrite (2.33a) in tensorial form and use the interference modulation tensor ΔM

$$\Delta X_{ij} = g^{\mathrm{p}}_{ijkl} \cdot \Delta M_{kl} \ . \tag{2.33b}$$

Here i, j, k, l stand for the spatial coordinates x, y, z and the Einstein sum convention is to be applied. In general, g^{p}_{ijkl} is a fourth-rank tensor. Note that the tensorial product in (2.33b) allows for a nonvanishing ΔX_{ij} even if $A_{\mathrm{a}} \perp A_{\mathrm{b}}$, i.e., vanishing intensity modulation. Such odd contributions to ΔX_{ij} are needed to account for polarization dependent interactions such as the dichroic bleaching of a dye to be discussed in Sect. 3.4.

The various physical mechanisms contributing to optical grating formation will be discussed in detail in Chap. 3, while here the framework for a formal description is to be worked out.

2.3.2 Optical Gratings

The material excitation, in general, couples to the refractive index and absorption coefficient which then also exhibit a grating-like modulation with amplitudes $\Delta n(\lambda_{\mathrm{C}})$ and $\Delta K(\lambda_{\mathrm{C}})$. Both Δn and ΔK are, of course, functions of the probe wavelength λ_{C}. The refractive-index modulation caused by a temperature grating, for example, is $\Delta n = (\partial n/\partial T) \cdot \Delta T$, where ΔT is the temperature amplitude and $(\partial n/\partial T)$ is the thermo-optic coefficient. $(\partial n/\partial T)$ has to be taken with certain constraints to be discussed in Sect. 3.7. Generally speaking, any modulation of a material property with amplitude ΔX inside a medium will be accompanied by an optical grating with amplitudes

$$\Delta n = (\partial n/\partial X)\Delta X \ , \tag{2.34a}$$

$$\Delta K = (\partial K/\partial X)\Delta X \ , \tag{2.34b}$$

where the tensor character of ΔX has been ignored for the moment. Quite frequently, one of the coupling constants $(\partial n/\partial X)$, $(\partial K/\partial X)$ is very small; the grating is then either of the phase or the amplitude type.

Instead of using two optical parameters, namely the absorption coefficient K and refractive index n, it is often convenient to combine these to a complex refractive index

$$\tilde{n} = n + iK/2\,k_c \ , \tag{2.35a}$$

$$\Delta\tilde{n} = \Delta n + i\Delta K/2\,k_c \ , \tag{2.35b}$$

where k_c is the absolute value of the wave vector of the light for which the optical properties are measured. Equations (2.34 and 35) can be combined into the more general expression

$$\Delta\tilde{n} = (\partial\tilde{n}/\partial X)\Delta X \ . \tag{2.35c}$$

2.3.3 Tensor Gratings

The complex refractive index is related to the complex optical frequency dielectric constant ε and the susceptibility χ by

$$\tilde{n}^2 = \varepsilon = 1 + \chi \ , \tag{2.36a}$$

$$\Delta\tilde{n} \approx \Delta\varepsilon/2\varepsilon^{1/2} = \Delta\chi/2(1+\chi)^{1/2} \ . \tag{2.36b}$$

Thus, an optical grating corresponds to a spatial modulation of any of the quantities $\tilde{n}$, ε, or χ.

Note that ε and χ are tensors, in general, while $\tilde{n}$ is not. Therefore, if anisotropic interaction is important, susceptibilities should be used for general description. Specifically, the susceptibility component χ_{ij} connects the electric-field component A_j with the polarization component P_i (where again $i,j = x, y, z$) by means of

$$P_i = \varepsilon_0\chi_{ij}A_j \ , \tag{2.37a}$$

$$\Delta P_i = \varepsilon_0\Delta\chi_{ij}A_j \ . \tag{2.37b}$$

The tensorial character of $\Delta\chi_{ij}$ includes induced birefringence and dichroism, i.e.

a polarization-dependent refractive index and absorption coefficient. Because both ΔX and $\Delta \chi$ are generally tensors of rank 2, the coupling constant between them is of rank 4 namely

$$\Delta \chi_{ij} = (\partial \chi_{ij}/\partial X_{kl})\, \Delta X_{kl} \; . \tag{2.38}$$

χ_{ij}, $\Delta \chi_{ij}$, and $(\partial \chi_{ij}/\partial X_{kl})$ are generally complex numbers to account for both index of refraction and absorption. Equation (2.38) shows that the anisotropy of $\Delta \chi_{ij}$ may be either due to the sample medium itself (crystals, external forces, flow) or be induced by the grating formation process. Even in an isotropic solid, thermal expansion, for instance, creates anisotropy, namely strain along the direction of $\boldsymbol{q}$ but stress in the planes perpendicular to $\boldsymbol{q}$. Equation (2.38) can be combined with (2.33b) to connect the optical grating amplitude directly with the pump field under stationary conditions

$$\Delta \chi_{ij} = f_{ijkl}\, \Delta M_{kl} \quad \text{where} \tag{2.39}$$

$$f_{ijkl} \equiv (\partial \chi_{ij}/\partial X_{k'l'}) \cdot g_{k'l'kl} \; . \tag{2.40}$$

2.3.4 Grating Excitation with Short Pulses

With short pulses, transient effects prevail. Then g^{p}_{ijkl} in (2.33b) and f_{ijkl} in (2.40) have to be replaced by appropriate functionals which will be discussed in context with the detailed discussion in Chap. 3 and specific experiments in Chap. 5.

Further, a coherent optical pulse creates excited atoms, the electronic wavefunctions of which are also coherent for some time. As a consequence, they behave like a set of elementary dipoles oscillating in phase at the excitation frequency. If the dephasing time is sufficiently large, it is possible to form an excitation grating even without overlap of pump pulses, i.e., with a vanishing ΔM. The grating is set up by interaction of the second pump pulse with the oscillating dipoles in this case (Sect. 3.3).

2.4 Detection of Gratings

Light-induced gratings can be probed with a third laser beam with intensity I_C, often of a frequency ν_C, different from that of the excitation beams. The grating diffracts some of the probe light into various directions. When the probe beam traverses the grating area, it creates a periodic polarization

$$\Delta P_{\text{C},i} = \Delta \chi_{ij} A_{\text{C},j} \; . \tag{2.41}$$

The radiation emitted from different parts of the polarization grating interferes constructively only in certain directions with regard to grating orientation and probe beam direction, i.e., the directions of q and the probe beam wave vector k_C.

The amplitudes A_m or intensities $I_m = \varepsilon_0 n c A_m^2/2$ of the different orders $m = \pm 1, \pm 2, \ldots$ are a measure for the modulation of the complex refractive index and the material excitation. The diffraction process is frequently called scattering in analogy to classical light scattering from random fluctuations.

The characteristics of the diffraction process is strongly dependent on the sample thickness d. If d is of the order of the grating period Λ or smaller, the grating is called thin, otherwise it is called thick. Detailed criteria for thin and thick gratings will be given in Chap. 4.

2.4.1 Thin Gratings

The Fourier transform of a thin grating is not an isolated spike on the k_x axis, namely $q_x = q = |q|$, but contains k_z contributions of the order d^{-1} caused by the finite thickness of the sample. As a result, constructive interference is possible at arbitrary directions of the probing beam. Further, several orders of diffraction can be observed in general (Fig. 2.5a). The familiar laws of diffraction yield for the directions ϕ_m of the different orders m

$$\Lambda \left[\sin(\phi_m + \alpha) - \sin \alpha \right] = m \lambda_C, \quad m = 0, \pm 1, \pm 2, \ldots, \quad (2.42)$$

where α is the angle of incidence. When α and ϕ_m are sufficiently small, the constant angular separation ϕ between the various diffracted beams is

$$\phi = \phi_{m+1} - \phi_m \approx \lambda_C / \Lambda . \quad (2.43)$$

In vectorial form, (2.42) means that the component k_{mx} of the wave vector k_m of the diffracted wave is given by the corresponding component k_{Cx} of the wave

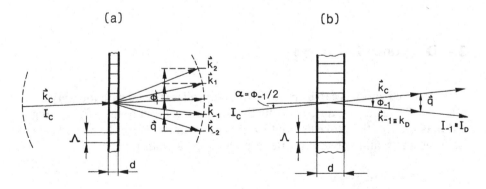

Fig. 2.5a,b. Characteristics of diffraction at **(a)** thin and **(b)** thick gratings

vector k_C of the incident probe beam plus some integer multiple of the grating constant q

$$k_{mx}=k_x+mq, \quad m=\pm1, \pm2, \ldots . \qquad (2.44)$$

The absolute values of the wave vectors of the probe and diffracted beams are equal ($k=k_m$) since there is no frequency change. For a stationary grating, the vector q is defined only up to a factor of ±1 according to (2.8). In (2.44) however, it is sufficient to consider only one direction q (Fig. 2.2) since m can be both positive or negative.

2.4.2 Thick Gratings

The Fourier transform of a thick grating is dominated by $q_x=|q|$ with negligible contributions in the other directions. As a consequence, thick gratings can be probed efficiently only if the Bragg condition

$$k_m-k_C=mq, \quad m=\pm1, \pm2, \ldots \qquad (2.45)$$

is obeyed. Figure 2.6a shows the highly symmetric geometrical arrangement while Fig. 2.6b shows the corresponding k vector diagram. In contrast to (2.44), the Bragg condition (2.45) determines completely the wave vector k_m of the diffracted wave and not only its x-component. In addition, the absolute values of wave vectors k_C and k_m are equal. This imposes a restriction on the wave vector k_C of the probe beam: the angle of incidence has to be half the diffraction angle:

$$\sin|\alpha|=\sin|\phi_m/2|=m\,\frac{\lambda_C}{2\Lambda}, \quad m=\pm1, \pm2, \pm3\ldots . \qquad (2.46)$$

Unless otherwise mentioned, we consider first-order diffraction in our further discussions involving thick gratings. In this context, the following notation will

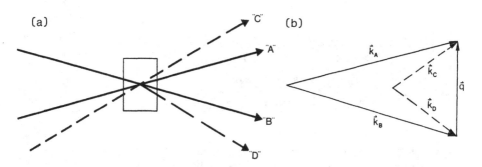

Fig. 2.6a,b. The highly symmetric arrangement of pump, probe and (first-order) diffracted beams illustrating the beam notation by subscripts A, B, C, D. (a) ray directions, (b) k and q vector relation

be used from now on: The subscript D stands for the diffracted (first-order) beam; thus, the sequence of subscripts A, B, C, D accounts for the four relevant beams in this book, the two pump beams, the probe and the (first-order) diffracted beam (Fig. 2.6).

The case $\Lambda = \lambda_C/2n$ is of special interest. The diffraction angle becomes $\phi_d = 180°$ which means that the diffracted beam has the opposite direction as the incident beam.

2.4.3 Degenerate Interaction and Self-Diffraction

We consider the case $k_C = k_A$ or k_B in this section, i.e., the probe beam has the same direction and wavelength as the pump radiation. The pump and probe/signal beams can be distinguished from each other in this case by different polarizations or by temporal delay. Otherwise, the pump and probe beams

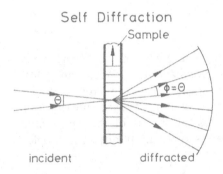

Fig. 2.7. Self-diffraction: the incident light waves produce the grating and are also diffracted

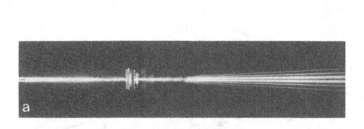

Fig. 2.8a, b. Self-diffraction: (**a**) experimental arrangement with light rays made visible by scattering at dust particles. Grating excitation was performed with a frequency-doubled Nd:YAG laser. A dye solution was used as grating material. (**b**) Diffraction pattern from the self-diffraction experimental of (**a**)

become identical. The existence of a grating can still be probed if the pump beams are of unequal intensity: the weaker may then become amplified at the expense of the stronger one. This is equivalent to the amplifier experiment in stimulated scattering mentioned in Sect. 1.2.2. Finally, if the grating amplitude is large enough, higher-order diffracted beams may indicate the existence of a grating. This phenomenon is called self-diffraction.

Self-diffraction is best observed with a thin grating and sufficiently small sample absorption (Fig. 2.7). The experimental observation of the self-diffraction effect is straighforward, as is obvious from Fig. 2.8, but its theoretical description is more complicated than in the case of separate excitation and probe beams.

Because of its experimental simplicity, self-diffraction can be used favorably for qualitative investigations of material properties. Note that the first-order diffracted beam of the lower pump beam coincides with the upper pump beam behind the sample, and vice versa.

2.4.4 Direct Detection

The amplitude diffracted into the first order ($m = \pm 1$) from a grating is to a first approximation, proportional to the (complex) refractive-index or susceptibility modulation amplitude $\Delta \tilde{n}$ or $\Delta \chi$ which, in turn, is generally proportional to the modulation amplitude ΔX of the respective material excitation, as discussed in Sect. 2.3.2. For ideal plane-wave interaction, i.e., spatially constant intensity and grating amplitudes, the normalized intensity of the first-order diffracted beam I_D/I_C is

$$I_D/I_C = \eta = \left| \frac{\pi \Delta \tilde{n} d}{\lambda_C} \right|^2 = \left(\frac{\pi \Delta n d}{\lambda_C} \right)^2 + \left(\frac{\Delta K d}{4} \right)^2 . \tag{2.47}$$

This formula for the diffraction efficiency η is valid for gratings with sufficiently small $|\Delta \tilde{n}|$ and also low absorption in the material, i.e. $Kd \ll 1$. The diffraction efficiency for arbitrary values of $|\Delta \tilde{n}|$ is different for thin and thick gratings to be discussed in Chap. 4. For beams of finite width, the intensities in (2.47) can be replaced by the respective ratio of the light fluxes (P_D/P_C).

Very small refractive-index changes Δn and optical path changes Δnd can be measured by diffraction. A diffraction efficiency of 10^{-5} is easily detected, corresponding to an optical path change $|\Delta \tilde{n} d| \approx \lambda/1000$. The phase shift is thus measured with interferometric sensitivity.

Amplitude ($\Delta n = 0$) and phase ($\Delta K = 0$) gratings may be distinguished by illumination with a parallel beam and observation of the self-images which appear at different distances behind the grating [2.6], or by heterodyne detection.

2.4.5 Heterodyne Detection

If the diffraction intensity is very small, the unavoidable background of stray light at the receiver may be more intense than the scattered light. Under such circumstances, the signal, recognizable by its time dependence, not only rides on a more-or-less sizeable background, but also depends on the relative phase between diffracted and stray light. It can be of advantage in such a situation to provide a fourth beam as a reference (R). The reference beam has to be made parallel to the diffracted one, i.e., $k_R = k_D$, and coincident at the detector (Fig. 2.9). The material excitation can now be recorded phase-sensitively by means of heterodyne detection. The experimental setup is characterized by a symmetrical arrangement of the "probe" and "reference" = "diffracted" beams, as sketched in Fig. 2.10.

Again assuming ideal plane-wave superposition, the normalized first-order diffracted heterodyne signal I_H/I_C at the detector is

$$I_H/I_C = I_R/I_C + 2 \frac{\sqrt{I_R I_D}}{I_C} \cos \psi_{RD} + I_D/I_C . \qquad (2.48)$$

The first term is constant but contributes to the noise [2.7]. The third term is given by (2.47) but is unimportant since $I_R \gg I_D$. The relevant information is contained in the second term: the square-root means that the diffracted-light amplitude $|E_D|$ and its phase relative to the reference beam ψ_{RD} determine the heterodyne signal.

Heterodyne detection hence provides $\Delta\tilde{n}$ rather than $|\Delta\tilde{n}|^2$ obtained from direct detection, as discussed previously. This is a great advantage when $|\Delta\tilde{n}| \ll 1$, and lowers the detection limit to the order of $|\Delta\tilde{n}| \simeq 10^{-16}$ [2.8]; it also provides information on real and imaginary parts of $\Delta\tilde{n}$, in particular on the signs of Δn and ΔK.

In order to detect very small signals, it may be necessary to integrate over extended periods of time. Phase instability due to (slow) thermal drifts etc., may then become a problem because a drift of $\Delta\psi_{RD} = \pi$ reverses the signal. Thus, during signal averaging, one may first observe a signal coming out of noise but vanish again before a sufficient signal-to-noise ratio has been achieved.

In such a situation, active phase stabilization may be useful. One possibility is to split off part of the pump, probe, and reference beams and to bring them to interfere outside of the sample (Fig. 2.10a). A transmission grating placed in the interference zone and rotated to match the $|q|$ of the interfering beams, transmits them only if the interference maxima coincide with the transparent stripes of the transmission grating.

This scheme allows detection of the phase of an interference pattern with regard to the fixed external grating. It can be locked to any desired value by means of standard electronic equipment. By extension to both interference patterns, their relative phase can also be kept at a fixed value. In doing so, the above-mentioned refractive index changes of 10^{-16} were detected by *Pohl* [2.8].

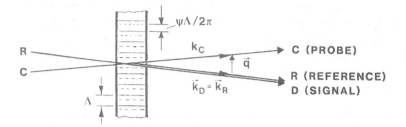

Fig. 2.9. Heterodyne detection of diffracted light. C and R beams form a second interference pattern (– – –) phase shifted by the amount ψ_{RD} with regard to the pump grating (——)

(a)

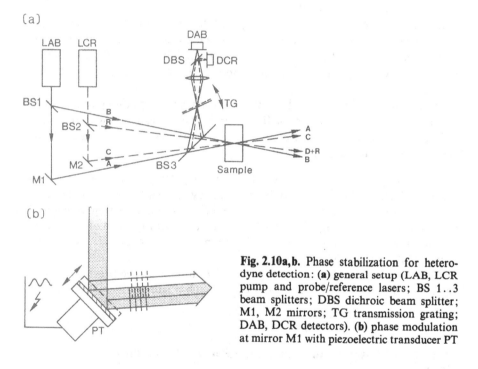

(b)

Fig. 2.10a,b. Phase stabilization for heterodyne detection: (a) general setup (LAB, LCR pump and probe/reference lasers; BS 1..3 beam splitters; DBS dichroic beam splitter; M1, M2 mirrors; TG transmission grating; DAB, DCR detectors). (b) phase modulation at mirror M1 with piezoelectric transducer PT

We return once more to (2.48). There is a close similarity to the expression for pump beam interference, (2.17). Probe and reference beams do indeed form a second interference pattern with the same grating vector q as the pump beams. ψ_{RD} represents the phase difference between the two gratings. Of course, the amplitude of the second grating has to be much smaller than that of the first one, to avoid undesirable extra material excitation.

In the derivation of (2.48), equal polarizations of the diffracted and reference beams have been assumed. A generalization to nonparallel polarizations of these two beams would indeed not make much sense because common photodetectors are sensitive to intensities and not to field strengths.

The polarizations of probe and diffracted beams, however, will in general not be the same if $\Delta\chi_{ij}$ contains odd terms ($i \neq j$). If the reference beam was polarized parallel to the probe, the superposition of diffracted and reference beams results in a modulation depth smaller than expected from (2.48).

The phase sensitivity of the heterodyne signal can be exploited in several ways for detection processes, for instance:

1) The pump beams are stationary but their relative phase gets shifted periodically between 0 and π, e. g. by moving a mirror mounted on a piezoelectric translator (Fig. 2.10b). The pump-grating phase oscillates correspondingly, and so does the diffracted-beam phase. Superposition with the reference beam results in a periodic reversal of the signal. Lock-in detection provides increased discrimination against background as a result.

2) The relative phase of pump and probe/reference gratings for maximum signal provides information on the nature of the grating, i. e., the values of real and imaginary parts of the refractive index Δn which in turn yields information on the material excitation process and coupling parameters not easily accessible by other means.

3) Flow measurements: A short pump pulse writes a grating pattern which drifts away with the velocity of the flow; hence, it changes phase with regard to the interogating grating. The frequency of phase reversal is a measure for the flow velocity component in direction of q.

In the last example, the domain of transient grating techniques begins to overlap with the field of laser Doppler velocimetry, also called laser anemometry [2.9]. Instead of a transient grating, small material particles floating in the fluid are the source of scattering of a probe beam. Since the arrangement of particles is random, the detection process is more complicated requiring, as a rule, autocorrelation techniques [2.10].

Conventional laser anemometry, of course, fails if no seeds are available. Transient gratings can provide a way out in such a situation.

2.5 Experimental Geometries for Grating Production and Detection

Laser-induced gratings have been produced and detected with a variety of different arrangements, an overview of which will be presented in this section. Fig. 2.11a is a simple combination of Fig. 2.2 – grating production – and Fig. 2.5a – probing of a thin grating. First-order diffraction is detected.

The beam used to detect the grating may be derived from one of the incident beams (Fig. 2.11b, d–g). This method is particularly convenient with ultra-short laser pulses. By varying the delay time of the probing beam, transient grating effects can be investigated with picosecond or femtosecond resolution [2.11] (Fig. 2.11b).

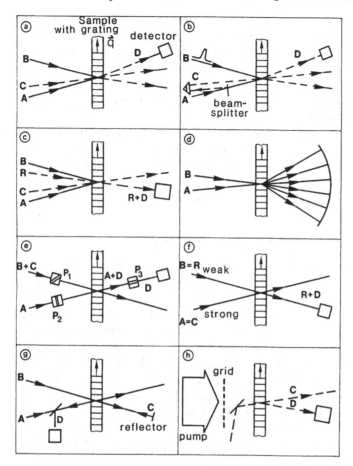

Fig. 2.11a–h. Arrangements for production and detection of laser induced gratings described in the text

Figure 2.11c sketches the schematics of a heterodyne arrangement [2.12]; similar to Fig. 2.11a, it is simply a combination of Figs. 2.2 and 2.9.

The self-diffraction arrangement of Fig. 2.7 is reproduced in Fig. 2.11d [2.13]. In Fig. 2.11e, polarization of the two pump beams differs by 45° [2.14]. If polarizer P_3 is set orthogonal to P_2, the diffraction from the upper pump beam will be observed without disturbance from the transmitted part of the lower one.

Figure 2.11f sketches an amplifier experiment. This type of interaction has been investigated in stimulated Rayleigh amplification experiments (Sect. 1.2.2) [2.15]. It is applied nowadays for image amplification by dynamic gratings in electro-optic crystals [2.16].

Figure 2.11g shows a frequently used four-wave setup: the probe beam is derived from one of the pump beams by a reflector behind the sample. This beam is diffracted into the direction opposite to the other primary (pump) beam which hence appears to be retroreflected although no physical mirror is present [2.17]. Care has to be taken when experiments of this type are evaluated, because a second grating is produced by interference of the beams with k_A and $-k_B$. Recently, the arrangement of Fig. 2.11g has attracted considerable interest, because the beam reflected into the direction of $-k_A$ shows phase conjugation or time reversal with respect to the incident beam with direction k_A (Sect. 6.9) and with respect to wave-front distortions.

Figure 2.11h shows an arrangement which is somewhat outside the framework envisaged up to now: the grating is created not by interference but simply by a grid-like aperture placed in front of a single pump beam with large diameter [2.18]. This arrangement is advantegeous for very large grating periods which would require impracticably small angles θ_p.

2.6 Spontaneous and Forced Light Scattering

The dynamic-grating technique is similar in many aspects to classical or spontaneous light scattering. In such a light-scattering experiment, the radiation diffracted by a Fourier component of a spontaneous statistical fluctuation of some material property is detected. The dynamic grating may be considered as just such a Fourier component, but with an artificially enhanced amplitude, forced by the incident light waves. Hence diffraction at laser-induced gratings is often called forced scattering. Historically this term was first used in connection with the special case of forced Rayleigh scattering (FRS, Sect. 5.2) [2.19]. In principle, all the information content of a classical light-scattering experiment can also be obtained from a transient-grating experiment, except for the statistical properties of random fluctuations; a transient-grating experiment, on the other hand, may also provide information on phase relationships which classical light scattering cannot.

Thermodynamic fluctuations, responsible for classical light scattering, are weak and at random. Transient-grating amplitudes can be much larger and are coherent. Therefore transient-grating investigations can provide highly increased detection sensitivity, as compared to the corresponding classical scattering experiment. The measurement of anisotropic heat conduction in liquid-crystal films or of second sound in a regular crystal and the so-called Mountain mode (Sect. 5.2) were possible only because of this high sensitivity.

As mentioned before, transient gratings also allow the investigation of microscopic excited states of the material with energies far above the thermal energy $k_B T$. The corresponding states are not thermally occupied; hence there is no correspondence to these contributions in classical light-scattering.

2.7 Transient Gratings in the Framework of Nonlinear Optics

The processes of grating creation and detection discussed so far involves four light waves in general: two pump beams at frequency ω_p, the probe beam, and the diffracted beam, both at frequency $\omega_C = \omega_D$. Since the four beams do not linearly superimpose, but influence one another, the grating generation and detection scheme falls into the regime of nonlinear optics.

The general approach to describe nonlinear optical interactions, which has now become the classical one, was introduced by Bloembergen and his co-workers in 1962 [2.20]. They started with the local, time-dependent polarization $P(r, t)$ which, in general, is a functional $f(E)$ of the local and time-dependent electrical field $E(r, t)$. The functional character (involving temporal and spatial operators) is of relevance mostly for transient effects (Sect. 2.3.1); in the steady state and assuming local response of the medium, $f(E)$ becomes a function of E.

The steady-state function $f(E)$ can be expanded into a power series in E. The right-hand side of the linear relation (2.37a) $P_i = \varepsilon_0 \chi_{ij} A_j$ is the first term of this series; at high fields, higher-order terms in A_j have to be considered, giving rise to nonlinear polarization terms P_i^{NL}.

To find the interaction between a certain number of light waves, Maxwell's equations have to be solved using the new expression for the polarization. This cannot be done in a general way; however, since the higher-order contributions to $P(r, t)$ are normally small, it is possible to consider separately the different light beams with frequencies $\omega_1, \omega_2, \omega_3 \ldots$ and wave-vectors $k_1, k_2, k_3 \ldots$ The respective higher-order terms can then be introduced as source terms into the wave equation. It is convenient further, to go from the instantaneous and local field E and polarization P to the amplitudes $A(\omega_1, k_1)$, $A(\omega_2, k_2)$, $A(\ldots)$ and $P^{NL}(\ldots)$ of the different waves involved.

In order to describe a certain source term, only those combinations of amplitudes are allowed for which the frequencies and wave vectors add up to the desired values of the source term. A careful book-keeping of vector and tensor components with indices $i, j, k \ldots$ (each index can have the values x, y, z) is required to properly account for anisotropic interaction. The resulting general and basic expression for a source term at a frequency combination $\pm \omega_1 \pm \omega_2 \pm \omega_3$ is

$$P_i^{NL}(\pm \omega_1 \pm \omega_2 \pm \omega_3 \ldots; \pm k_1 \pm k_2 \pm k_3 \ldots)$$

$$= \varepsilon_0 [\chi_{ijk}^{(2)}(\ldots) A_j(\ldots) A_k(\ldots) + \chi_{ijkl}^{(3)} A_j(\ldots) A_k(\ldots) A_l(\ldots) + \ldots] \ . \quad (2.49)$$

Negative values of the frequencies in the combination $\pm \omega_1 \pm \omega_2 \pm \omega_3 \ldots$ are obtained using $A_1(-\omega_1, -k_1) = A^*(\omega_1, k_1)$ etc. The coefficients χ are the higher order or nonlinear susceptibilities. They are tensors of rank two, three, four, and so on. All the coefficients are functionals (transient effects) or functions (stationary conditions) of the frequencies and, in principle, of the k-vectors involved [abbreviated $(\ldots)$ in (2.49)]. It is assumed in this derivation that the material is spatially homogeneous.

We shall now apply the general expression (2.49) to the case of transient-grating creation and detection. Our interest focuses on the source term for the scattered beam $P_D(\omega_D, k_D)$ which will be denoted P_D, for simplicity.

The four electric-field amplitudes will be denoted as

$$A(\omega_p, k_A) \equiv A_A \; ,$$

$$A(\omega_p, k_B) \equiv A_B \; ,$$

$$A(\omega_C, k_C) \equiv A_C \; ,$$

$$A(\omega_D, k_D) \equiv A_D \; .$$

The second-order term in (2.49) involves the combination of two field amplitudes and hence the sum or difference of the available frequencies: $2\omega_p, \omega_p \pm \omega_C, 2\omega_C, 0$. Thus, this term cannot contribute at the frequency $\omega_D = \omega_C$. The same is true with the other even-order terms. Thus we are left to lowest order with the third-order term which provides the required combination $\omega_D = \omega_p - \omega_p + \omega_C = \omega_C$. Similarly, possible combinations at all odd-order terms exist, but higher-order terms are usually smaller. The source term with ω_D radiates a new wave A_D which builds up monotonically if the respective wave-vectors are equal:

$$k_D = k_A - k_B + k_C \; . \tag{2.50}$$

Equation (2.50) is called the phase-matching condition in nonlinear optics. It corresponds to the first-order Bragg-condition [(2.45) with $m = 1$] which connects the directions of the probe wave k_C and the wave k_D diffracted at a grating with a grating vector $q = k_A - k_B$.

If the phase-matching condition (2.50) is not fulfilled, a scattered wave can still be observed if the sample is sufficiently thin. This situation corresponds to diffraction at a thin grating.

Considering only terms radiating the frequency $\omega_C = \omega_D$, the basic equation (2.49) reduces to

$$P_{D,i} = \varepsilon_0 (\chi_{ij}^{(1)} A_{D,j} + \chi_{ijkl}^{(3)} A_{C,j} A_{A,k} A_{B,l}) \; . \tag{2.51}$$

All the nonlinear interactions of the grating generation and detection process are now expressed in one single quantity, the third-order susceptibility $\chi_{ijkl}^{(3)}$. Considering the different physical mechanisms leading to this quantity, it is not surprising that $\chi_{ijkl}^{(3)}$ is a complicated parameter: it is a tensor of rank four with complex components. In addition, the nonlinear susceptibility tensor depends on four frequencies and wave vectors. Our case, however, is partially degenerate in so far as $\omega_A = \omega_B = \omega_p$, $\omega_D = \omega_C$ or fully degenerate if all frequencies are equal.

Introducing the source term (2.51) into the wave equation, the diffracted field strength A_D can be calculated. Thus, it is possible to describe the directions and intensities of the diffracted waves by nonlinear optical interaction (degenerate four-wave mixing) of the pump and probe beams.

We shall now show that the general nonlinear-optical-source term (2.49) can be reduced to a change $\Delta\chi_{ij}$ of the linear susceptibility for the probe beam. In our case of degenerate four-wave mixing, the polarization of the diffracted beam can be written in analogy to (2.51) as

$$P_{D,i} = \varepsilon_0(\chi_{ij}A_{D,j} + \Delta\chi_{ij}A_{C,j}) \quad \text{with} \tag{2.52}$$

$$\Delta\chi_{ij} \equiv \chi_{ijkl}^{(3)}A_{A,k}A_{B,l} \ . \tag{2.53}$$

Comparison of (2.53) with (2.37b) shows that the terms $\Delta\chi_{ij}$ in the two equations are identical. In the case of an isotropic medium and $A_A \| A_B$, $\Delta\chi_{ij}$ becomes a scalar $\Delta\chi$ which can be expressed as a change $\Delta\tilde{n}$ of the refractive index according to (2.36b). Similarly, the nonlinear refractive index n_3 can be introduced converting (2.53) into

$$\Delta\tilde{n} \equiv n_3|A_A| \, |A_B| \tag{2.54}$$

In summary, the general approach of nonlinear optics comprises the transient-grating situation as a special case. The formal description by nonlinear susceptibilities does not provide new information on the material excitations involved, but is an elegant short-hand notation for the evaluation of electromagnetic-beam interaction. As the mechanisms of grating excitation are the primary objective of this volume, the long-hand notation previously introduced (Sect. 2.3) appears to be more appropriate here.

3. Mechanisms of Grating Formation and Grating Materials

In the following section, different mechanisms of light-induced changes of optical material properties are described. Often these changes are considered to be due to some specific material excitation (e.g., electron density, temperature). This is, however, a simplified description. In general, the material is excited in different ways and the various excitations are coupled. A separation of these excitation can be achieved, at last partially, by considering a short excitation pulse. Different excitations then develop at characteristic time ranges after the pulse.

3.1 Interaction Mechanisms: Real and Virtual Transitions

Ultraviolet, visible, and near-infrared light ($\lambda < 3$ μm) primarily interacts with the electrons of the matter it traverses; mid- and far-infrared radiation interacts with its phonons and other low-energy excitations.

The interaction may be a real transition or absorption process if the light frequency matches a resonance of the medium. An excited state of the medium is populated, and a photon is annihilated. After some time, a photon is reemitted, generally with different frequency and direction. It is also possible that the energy of the excited state is transferred into some other kind of excitation, e.g. heat.

If the frequency of the light is off resonance, the light field may also affect the material and therefore change itself. The underlying interaction process is often described as a virtual transition. This term implies a phenomenological and verbal description of the interaction effect which is explained in detail by quantum theory. To describe an off-resonance interaction, a virtual absorption process is assumed where an incident photon is annihilated, but another photon is created simultaneously. The new photon may be of lower, same or higher energy than the incident one; moreover, it may be emitted in the direction of the incident photon, or in a different direction.

There is no sharp distinction between the two types of interaction since any resonance is of finite width. For practical purposes, however, real absorption processes denote strong interactions changing appreciably the intensity of the incident radiation. Virtual processes, on the other hand, are weak interactions changing the intensity comparatively weakly. Nonresonant or virtual processes are the microscopic origin of, e.g., Raman and Brillouin scattering, the optical Kerr effect and various mixing effects due to nonlinear electronic polarizabilities. Raman and Brillouin scattering associated with travelling waves are of less interest here, as pointed out before; the other nonresonant processes cause quasistatic refractive index changes and therefore are of major importance here. However, most of the grating mechanisms discussed in the following section are due to resonant interactions.

We shall next discuss the sequence of excitations/de-excitations following the absorption of a laser pulse by an electronic transition. The corresponding grating structures will be discussed in detail in the subsequent sections.

3.2 The De-Excitation Sequence

An absorbed short laser pulse leaves the material system, in the first moment, in a coherently excited electronic state, i.e., the electrons oscillate with a constant phase relation to the light field. Grating effects, exploiting coherent excitation are described in Sect. 3.3. In general, coherence is lost rapidly, characterized by a time constant which is of the order of 10^{-15} to 10^{-7} s. The resulting, incoherently excited state population may be similarly short-lived, but may also persist up into the millisecond range and may produce population density gratings (Sect. 3.4).

In solids (semiconductors) with electrons and holes of high mobility, the population density grating corresponds to a free carrier grating (Sect. 3.5). In electrooptic crystals, free carriers move and build up space charges which enhance the grating effects via the electrooptic effect (Sect. 3.6).

Electrons decaying from the primarily excited state may populate intermediate electronic or other excited states or induce a chemical reaction. The resulting secondary structures may be longer lived or even (meta)-stable (photochromism). In the end, the absorbed energy gets thermalized and the population grating leaves behind a thermal grating (Sect. 3.7). This grating decays by heat diffusion, which typically takes 10^{-4} to 10^{-2} s. The temperature grating creates density, stress, and strain gratings by thermal expansion. In a fluid, convective velocity gratings, and in mixtures, concentration gratings can be induced in ternary processes. Decay times of 10^{-2} to 10 s are typical in this case (Sect. 3.8).

The different material excitations developing from exited electronic states are put together for an overview in Fig. 3.1. Similar sequences of grating excitations may be induced by infrared pump light which primarily excites vibrational states.

Contributions from the various types of gratings to the light scattering can be identified by their intensity, time dependence, and polarization.

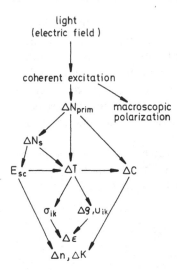

Fig. 3.1. Possible sequences of material excitations produced by a short light pulse. [N: Primary (secondary) population density of excited electronic levels, E_{sc}: space charge field, ΔT: temperature change, ΔC: concentration change, $\Delta \varrho$: density change, u_{ik}: strain, σ_{ik}: stress, $\Delta \varepsilon$: change of permitivity, Δn: refractive index change, and ΔK: absorption change]

3.3 Gratings Produced by Coherent Interaction of Atomic States

The primary effect during the interaction of light with matter is the forced oscillation of the electrons. If a light pulse is suddenly turned on, the electrons start to oscillate coherently with the light field strength. After some time T_2, the coherence gets lost and secondary material excitations build-up. In the following an experiment is discussed [3.1–3] where a grating is produced by the primary coherent interaction. Interestingly such a grating can be excited by two excitation pulses which are delayed by a time larger than the pulse width. Superposition of such two light beams does not result in an interference pattern, i.e. a spatially modulated light intensity. Nevertheless by using coherent interaction, it is possible that such delayed light beams produce a spatially modulated material excitation, i.e. a grating. The following discussion of gratings produced by coherent interaction emphasizes this interesting possibility.

Grating formation of two strongly delayed excitation pulses is possible if the phase of the first light wave packet is stored by the material and preserved until the second light pulse appears. Phase storage is achieved because the light wave produces electrons or atomic dipoles, which oscillate coherently with the light wave. The same is true for the macroscopic polarization given by the dipole density. The phase of a light wave consequently can be stored by the polarization of a material.

According to quantum mechanics, an atomic dipole moment exists if the atom can be described by a coherent superposition of two energy states (e.g.,

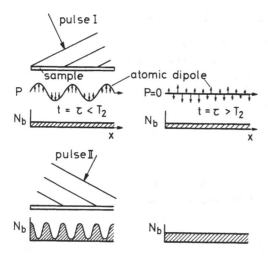

Fig. 3.2. Interaction of two light pulses (delay time τ) with a two-level system (polarization decay time T_2). $\tau < T_2$: The first pulse I produces an excited state population N_b and, in addition atomic dipoles with density P (polarization). P oscillates in phase with respect to the light field also after pulse termination. The second pulse II interacts with the atomic dipoles and changes N_b depending on the relative phase of the light wave II and the initial polarization p at $t = \tau$. N_b gets spatially modulated. $\tau > T_2$: The atomic dipoles produced by pulse I are out of phase at the beginning of pulse II; therefore $P = 0$ at $t = \tau$. The final excited state population N_b is spatially constant

ground and excited states). The phase of a light wave producing transitions between these two states is stored in the relative phase of the wave functions of the two states. After the light wave is switched off, the polarization decays in a time T_2 which is often much shorter than the decay time T_1 of the excited state population of the material. The polarization decay time T_2 is also called phase relaxation time because the polarization decays by dephasing of the atomic dipoles due to interactions with other excitations (e. g., thermal phonons). At low temperatures, the phase relaxation time can be quite long, e. g. in ruby at 2 K a value $T_2 = 10^{-7}$ s is observed.

For grating production (Fig. 3.2) the material is excited by a first pulse with a width t_p. The second pulse with different direction and delayed by a time interval $\tau > t_p$ interacts with the polarization left by the first pulse and produces a spatially modulated population density when the condition

$$\tau < T_2 \tag{3.1}$$

is satisfied. The population density grating can be detected by diffraction of a third light beam as usual.

3.3.1 Calculation of the Population-Density Modulation

The material is described [3.4] as a two-level system with a transition energy $\hbar\omega_0$. The total number density of atomic systems is N. The density in the lower level is N_a, while in the upper level it is N_b. The density difference $\gamma = N_a - N_b$ changes due to the influence of a light field $E = E(t)$:

$$\frac{d\gamma}{dt} + \frac{1}{T_1}(\gamma - N) = -\frac{2E}{\hbar\omega_0}\frac{dP}{dt} \ . \tag{3.2}$$

The polarization $P = P(t)$ is given [3.4] by the following equation resembling the equation of motion of a forced harmonic oscillator:

$$\frac{d^2 P}{dt^2} + \frac{2}{T_2}\frac{dP}{dt} + \omega_0^2 P = \frac{2\omega_0}{\hbar}\omega^2\gamma E \ . \tag{3.3}$$

Here μ is the dipole matrix element (without orientational averaging) for the transition between the two energy levels.

For further discussion of (3.2, 3) the slowly varying amplitude approximation is used

$$E = \tfrac{1}{2}\bar{E}(t)\exp\left[i(\omega_1 t - \boldsymbol{k}\cdot\boldsymbol{r})\right] + \text{c.c.} \ , \tag{3.4}$$

$$P = \tfrac{1}{2}\bar{P}(t)\exp\left[i(\omega_1 t - \boldsymbol{k}\cdot\boldsymbol{r})\right] + \text{c.c.} \ , \tag{3.5}$$

which results in the following approximate equations for the amplitudes $\bar{E}(t)$ and $\bar{P}(t)$ of the light field and polarization:

$$\frac{d\gamma}{dt} + \frac{1}{T_1}(\gamma - N) = \frac{i}{2\hbar}\left[\bar{E}(t)\bar{P}^*(t) - \bar{E}^*(t)\bar{P}(t)\right] \ , \tag{3.6}$$

$$\frac{d\bar{P}(t)}{dt} + \left[\frac{1}{T_2} + i(\omega_1 - \omega_0)\right]\bar{P}(t) = -\frac{i}{\hbar}\mu^2\gamma\bar{E}(t) \ . \tag{3.7}$$

For times $t \gg T_2$, Eq. (3.7) can be approximated by the steady-state solution $P \approx \varepsilon_0\chi E$ with the susceptibility χ given by (3.33). Insertion of this approximation into (3.6) gives a rate equation, see (3.36), for the population difference γ produced by the light intensity proportional to $|E|^2$. Thus, for times $t \gg T_2$, the basic equations used for the description of population density gratings in Sect. 3.4 are obtained.

In the following, (3.6, 7) are discussed for times t not too large compared to T_2. For description of grating formation by two delayed optical pulses, the field strength is assumed to have a rectangular pulse envelope:

$$E_{I,II}(t) = \tfrac{1}{2}\,\bar{E}_{I,II}(t)\,\exp\left[i\left(\omega_1 t - k_{I,II}\cdot r\right)\right] + \text{c.c.} \tag{3.8}$$

with

$$\bar{E}_I(t) = \begin{cases} \bar{E}, & 0 \le t \le t_p \\ 0, & t < 0,\ t > t_p \end{cases} \tag{3.9}$$

and

$$\bar{E}_{II}(t) = \begin{cases} \bar{E}, & \tau \le t \le \tau + t_p \\ 0, & t < \tau,\ t > \tau + t_p\ . \end{cases} \tag{3.10}$$

Here $\bar{E}_{I,II}$ and $k_{I,II}$ are the electric field amplitudes and wave-vectors of the two pulses. The directions of k_I and k_{II} are different. To simplify the discussion, only the case $\omega_1 = \omega_0 = \omega$ will be treated. Because the amplitude $\bar{E}_{I,II}(t)$ has constant values ($\bar{E}$ or zero) during the different time intervals, it can be treated as a constant parameter. The solution is further simplified by splitting the polarization amplitude into the real and imaginary parts $\bar{P}_1$, $\bar{P}_2$:

$$\bar{P}(t) = \bar{P}_1(t) + i\bar{P}_2(t)\ . \tag{3.11}$$

Because $t_p \ll T_1, T_2$, the change of the population difference γ and the polarization $\bar{P}(t)$ during a light pulse is given from (3.6, 7 and 11) by

$$\frac{d\gamma}{dt} = \frac{1}{\hbar}\,\bar{E}\bar{P}_2 \tag{3.12}$$

$$\frac{d\bar{P}_1(t)}{dt} = 0 \tag{3.13}$$

$$\frac{d\bar{P}_2(t)}{dt} = -\frac{\mu^2 \gamma \bar{E}}{\hbar}\ , \tag{3.14}$$

which leads with the use of (3.12) to

$$\frac{d^2\bar{P}_2(t)}{dt} = -\Omega^2 \bar{P}_2(t) \quad \text{with}$$

$$\Omega = \frac{\mu\bar{E}}{\hbar}\ . \tag{3.15}$$

During the first light pulse, γ and P are calculated from (3.5, 12, 13, 15) considering $\gamma(0) = N$, $P(0) = 0$ to

$$\gamma(t) = N\cos\Omega t, \quad 0 \le t \le t_p\ , \tag{3.16}$$

$$P(t) = -\frac{i}{2}\,N\mu\sin\Omega t\,\exp\left[i\left(\omega t - k_I\cdot r\right)\right] + \text{c.c.}\ . \tag{3.17}$$

Equation (3.17) shows that the phase of the first light wave is stored in the polarization. After the end of the first pulse, γ and P decay according to (3.2, 3), with $E=0$, as

$$\gamma(t)=N(\cos\Omega t_{\mathrm p}-1)\exp(-t/T_1)+N, \qquad t_{\mathrm p}<t<\tau \tag{3.18}$$

$$P(t)=-\frac{\mathrm i}{2}N\mu\sin(\Omega t_{\mathrm p})\exp(-t/T_2)\exp(\mathrm i(\omega t-\boldsymbol k_I\cdot\boldsymbol r)]+\text{c.c.} \tag{3.19}$$

In the experiment (Sect. 3.3.2) the pulse delay τ is small compared to T_1 so that the decay in (3.18) can be neglected.

During the second light pulse, γ and P develop from the initial values in the form

$$\gamma(\tau)=N\cos\Omega t_{\mathrm p} , \tag{3.20}$$

$$P(\tau)=-\frac{\mathrm i}{2}N\mu\sin(\Omega t_{\mathrm p})\exp(-\tau/T_2)\exp[\mathrm i(\omega\tau-\boldsymbol k_I\cdot\boldsymbol r)]+\text{c.c.} \tag{3.21}$$

The second light pulse now interacts with a polarization which is still in phase with the first light pulse. Because the wave-vector $\boldsymbol k_{II}$ of the second light pulse is different from $\boldsymbol k_I$, also the wave-vector of the polarization wave $P(t)$ for $t\geq\tau$ is written as

$$P(t)=\tfrac12\,\bar P(t)\exp[\mathrm i(\omega t-\boldsymbol k_{II}\cdot\boldsymbol r)]+\text{c.c.} \tag{3.22}$$

Comparison of the two preceeding equations, i.e. $P(t=\tau)=P(\tau)$, yields the following value of the polarization amplitude $\bar P(t)$ at $t=\tau$

$$\bar P(\tau)=-\mathrm iN\mu\sin(\Omega t_{\mathrm p})\exp(-\tau/T_2)\exp[\mathrm i(\boldsymbol k_{II}-\boldsymbol k_I)\cdot\boldsymbol r] , \tag{3.23}$$

$$\bar P_2(\tau)=-N\mu\sin(\Omega t_{\mathrm p})\exp(-\tau/T_2)\cos(\boldsymbol k_{II}-\boldsymbol k_I)\cdot\boldsymbol r . \tag{3.24}$$

The solution of (3.12, 13, 15) with these initial values $\gamma(\tau)$, $\bar P_2(\tau)$ is

$$\gamma(t)=\gamma(\tau)\cos\Omega(t-\tau)+\frac{\bar P_2(\tau)}{\mu}\sin\Omega(t-\tau), \qquad \tau<t<\tau+t_{\mathrm p} , \tag{3.25}$$

$$\bar P_2(t)=-\mu\gamma(\tau)\sin\Omega(t-\tau)+\bar P_2(\tau)\cos\Omega(t-\tau) . \tag{3.26}$$

At the end of the second light pulse, the population difference is given by:

$$\gamma(\tau+t_{\mathrm p})=N[\cos^2(\Omega t_{\mathrm p})-\sin^2(\Omega t_{\mathrm p})\exp(-\tau/T_2)\cos(\boldsymbol k_I-\boldsymbol k_{II})\cdot\boldsymbol r] . \tag{3.27}$$

The corresponding polarization P is not of interest here and is therefore not quoted. After the end of the second light pulse, the population difference γ decays according to (3.2) with the decay time T_1.

Equation (3.27) shows that, during the second light pulse, the population difference gets modulated with a spatial period $\Lambda = 2\pi/|k_I - k_{II}|$. A grating is therefore produced even though no interference of the two excitation pulses has taken place.

3.3.2 Experiment

The experimental setup [3.2, 3] is shown in Fig. 3.3. A 1 MW ruby laser and a beam splitter BS produce two pulsed light beams *I*, *II* with $t_p \approx 10$ ns. Beam *II* is delayed by an optical delay line ODL for about $\tau = 50$ ns. The two beams are incident on the ruby sample at an angle of 1° corresponding to different wave vectors k_I and k_{II} and produce a grating with a spacing $2\pi/|k_{II} - k_I|$. For clarity of presentation, the angle of 1° has been greatly exaggerated in the drawing. The grating is detected by diffraction of a third beam with $k_{III} = -k_{II}$ produced by reflection at mirror M. The Bragg condition (or phase-matching condition) gives the direction of the diffracted beam as $k = k_{III} + k_{II} - k_I = -k_I$. This beam is detected with a semitransparent mirror, photomultiplier and oscilloscope. The ruby sample with 0.05 % Cr^{3+}-concentration was cut perpendicular to the *C*-axis and had a thickness of 1.5 mm. Cooling to 2.2 K increases the phase relaxation time to 10^{-7} s. To compensate the accompanying shift of the absorption line center, the ruby laser rod had also to be cooled down to 80 K. In addition, a magnetic field was applied for fine-tuning. When the laser is cooled to 80 K, the 6933.97 Å laser line [the transition $^4A_2$ ($M_s = \pm 3/2) \rightarrow {}^2E(E)$ at 77 K] is at resonance with the 6943 Å absorption line of the sample [transition $^4A_2$ ($M_s = \pm 1/2) \rightarrow {}^2E(\bar{E})$ at 2.2 K]. All the beams were polarized perpendicular to the plane of Fig. 3.3 and Helmholtz's coils were used to produce a weak longitudinal field (0–250 G) about parallel to the crystal *C*-axis.

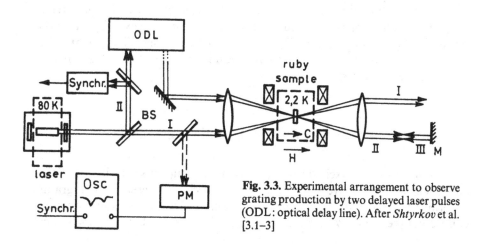

Fig. 3.3. Experimental arrangement to observe grating production by two delayed laser pulses (ODL: optical delay line). After *Shtyrkov* et al. [3.1–3]

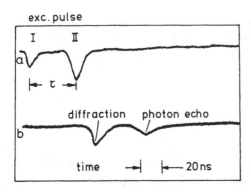

Fig. 3.4. Oscilloscope traces (a: delayed excitation pulses, b: diffracted pulse and photon echo) observed in the experimental arrangement according to Fig. 3.3 after *Shtyrkov* et al. [3.1–3]

Figure 3.4a shows an oscillogram of the pulses I and II with the delay τ. The pulse III is delayed 10 ns with respect to pulse II so that also the diffracted pulse in Fig. 3.4b appears with this delay. The second pulse on line b of the same oscillogram is a stimulated photon-echo signal [3.2, 5], which, as is explained in the reference, is formed with a delay τ after the third laser pulse. The appearance of the diffracted signal and the photon echo in the same direction is due to the fact that these phenomena are subject to the same phase-matching condition.

3.3.3 Related Effects

Optical effects due to coherently excited atoms have attracted a lot of interest in the last years [3.6]. All these phenomena (optical nutation, free induction decay, various types of photon echos, self-induced transparency) can be observed with resonantly excited particles before coherence is lost in the system. In grating experiments, atomic coherence effects must also be considered, as in the example of grating formation by two delayed light pulses which has just been discussed. Another interesting example is the appearance of grating echos [3.7] which are observed if population density gratings are produced by standing-wave excitation pulses.

In a more general sense, the interaction between light-waves with different propagation direction is observed in a laser-induced grating experiment. In this sense, nonlinear optical effects like photon echos with noncollinear light pulses [3.5, 8, 9] and resonant degenerate four-wave mixing [3.10–15] are closely related to grating experiments.

3.4 Population Density Gratings in Solids and Liquids

If an atomic system is excited from the ground to an upper state, the absorption coefficient and refractive index change, which can be observed in a grating experiment. In the following, we shall outline first some basic equations connecting the intensity of the exciting light field to the change of the optical properties which determine the diffraction efficiency of the corresponding grating. Afterwards, examples for population-density gratings in various materials will be given.

3.4.1 Absorption and Index Changes, Diffraction Efficiency

Simplified atomic systems are considered where only two electronic-energy levels become populated. These are represented by fictitious 2-level systems and by more realistic systems with 3 or more energy levels where only two levels are strongly populated. The change of the absorption coefficient can be calculated from the absorption cross-section σ_0 and σ_1 of the ground and excited states which are known experimentally for a number of materials. The absorption coefficient K is given by the population density N_0 and N_1 of the two states

$$K = \sigma_0 N_0 + \sigma_1 N_1 \ . \tag{3.28}$$

By optical excitation, the densities change to $N_0 - \Delta N$ and $N_1 + \Delta N$, producing a change of the absorption coefficient

$$\Delta K = -(\sigma_0 - \sigma_1)\Delta N \ . \tag{3.29}$$

The change of the absorption coefficient is accompanied by a change of the refractive index Δn. For many materials investigated, no measured values for Δn are available. It is then possible to estimate Δn by using the Kramers-Kronig relation [3.16].

$$\Delta n = \frac{1}{2\pi^2} \int_0^\infty \frac{\Delta K(\lambda)\, d\lambda'}{1 - (\lambda'/\lambda)^2} \ . \tag{3.30}$$

This formula is valid for small absorption coefficients (K and $\Delta K \ll 4\pi n/\lambda$) and small refractive index changes ($\Delta n \ll n$). $\lambda = 2\pi c/\omega$ is the vacuum wavelength.

The change of the absorption coefficient and the refractive index can be combined to express a change of the complex susceptibility. From

$$\chi = (n + iKc/2\omega)^2 - 1 \approx n^2 - 1 + iKnc/\omega$$

with $\quad K \ll 2\omega/c = 4\pi/\lambda, \quad$ one obtains

$$\Delta\chi \approx 2n\Delta n + i(nc/\omega)\,\Delta K \ . \tag{3.31}$$

With $\Delta\chi$, the diffraction efficiency η of a weak grating (thickness: d) is expressed as

$$\eta = \left(\frac{\pi\Delta n\, d}{\lambda}\right)^2 + \left(\frac{\Delta K d}{4}\right)^2 = \left(\frac{\pi d}{2n\lambda}\right)^2 |\Delta\chi|^2 \ . \tag{3.32}$$

The diffraction efficiency thus measures the change of the absolute value $|\varDelta\chi|$ of the complex susceptibility. This is a unique property of the grating method compared to transient absorption experiments where the change of the absorption coefficient $\varDelta K$ is observed.

Because optical excitation changes both the absorption constant and the refractive index, a population density grating is a mixed amplitude and phase grating. To obtain an understanding as to which contribution dominates, a simple 2-level-system with a transition frequency ω_0 and a halfwidth $1/T_2$ of the absorption curve is considered [3.4, 17]

$$\chi_{\text{model}} = \frac{|\mu|^2 (N_a - N_b)/\hbar\varepsilon_0}{(\omega_0 - \omega) + i/T_2} \ . \tag{3.33}$$

This equation can be obtained as the steady-state solution $P \approx \varepsilon_0 \chi E$ of (3.7).

The relative contribution of the amplitude (η_a) and phase (η_p) grating to the diffraction efficiency of this model system is given by

$$\frac{\eta_a}{\eta_p} = \left(\frac{\text{Im}\{\varDelta\chi\}}{\text{Re}\{\varDelta\chi\}}\right)^2 = \left(\frac{1/T_2}{\omega_0 - \omega}\right)^2 \ . \tag{3.34}$$

It follows that the amplitude grating dominates if the frequency ω of the probing radiation is near the transition frequency ω_0. If the difference $|\omega - \omega_0|$ is larger than the halfwidth $1/T_2$ of the absorption curve, a population density grating behaves like a phase grating.

For the evaluation of transient-grating experiments, it is desirable to relate the diffraction efficiency η to the intensity I or energy density E_p of the light inducing the grating. This can be done by recognizing that, according to (3.29–31) the change of the susceptibility is given by

$$\varDelta\chi = -(\alpha_0 - \alpha_1)\,\varDelta N \ . \tag{3.35}$$

Neglecting local field corrections, α_0 and α_1 are the polarizabilities of the ground and excited states of the atomic system. The change of the population density $\varDelta N = N_a - N_b$ can be obtained from a rate equation which is given in the simple case of *2-level-systems* with total density N by

$$\frac{\partial N_a}{\partial t} = \frac{N_b}{T_1} - \frac{\sigma I}{h\nu}(N_a - N_b), \quad N_a + N_b = N \ . \tag{3.36}$$

Here T_1 is the life-time of the upper level with population N_b, and σ the absorption cross section. The population densities are denoted by N_a and N_b instead of N_0 and N_1 to distinguish the 2-level-system from the 3-level-system to be discussed later. Equation (3.36) can be obtained from (3.6) as has been indicated already in Sect. 3.3. The rate equation (3.36) will not be discussed further but is given here mainly to demonstrate how rate equations can be connected to the more fundamental semi-classical description of light matter interaction outlined in Sect. 3.3.

Most of the materials investigated up to now (e.g., ruby, dye solutions) are better described as *3-level-systems*. Here absorption of photons (energy: hv) from the ground state (energy: E_0, population density: N_0) produces only a small population in the directly excited state with energy $E_0 + hv$. The directly excited state decays rapidly into some lower state with resultant population density N_1 and relatively long life time T_1. Because the population of the directly excited state is very small, the total density is given by $N \approx N_0 + N_1$. The optically induced population change in a *3-level-system* is obtained from the following rate equation

$$\frac{\partial N_0}{\partial t} = \frac{N_1}{T_1} - \frac{\sigma_0 I}{hv} N_0 \ . \tag{3.37}$$

The steady-state population change $\Delta N = N - N_0 = N_1$ is given by

$$\Delta N = N \frac{I/I_s}{1 + I/I_s} \quad \text{with} \quad I_s = hv/\sigma_0 T_1 \ . \tag{3.38}$$

If the material is excited by pulses short compared to the lifetime T_1, the population change ΔN_p is given by

$$\Delta N_p = N[1 - \exp(-\sigma_0 E_p/hv)] \quad \text{with} \quad E_p = \int I \, dt \ . \tag{3.39}$$

Population density gratings have been investigated experimentally in doped crystals (e.g., Cr ions in ruby, Nd ions in YAG) and also in dye solutions, as will be discussed in the following. Free-electron gratings in semiconductors and the spatial holes burnt into the upper-level population of laser materials can also be considered as population-density gratings. Because of their special properties, these types of gratings will be described later in Sects. 3.5 and 7.1.

3.4.2 Population Density Gratings in Ruby

The first laser-induced grating experiment in a solid has been performed with a ruby crystal pumped inside an optical resonator by the standing light wave of a ruby laser [1.10]. A spatially periodic population density is produced in the ^2E states of the Cr^{3+} ions in the crystal (Fig. 3.5) by absorption of the 694 nm ruby laser radiation from the $^4A_2$ ground state to the $\bar{E}$ excited state which is in thermal equilibrium with the $2\bar{A}$ state. The absorption coefficients K_0, K_1 of the ground and excited states [3.22–26] are shown in Fig. 3.6. The change of the refractive index due to optical pumping has been calculated, up to now, only in the vicinity of the 694 nm laser line on the basis of the known ground and excited states absorption spectra and (3.30) [3.26], see also [3.27–30]. These results are not

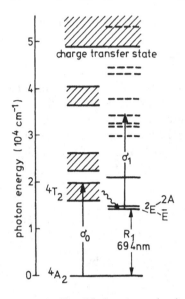

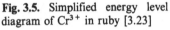

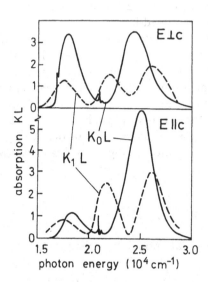

Fig. 3.5. Simplified energy level diagram of Cr^{3+} in ruby [3.23]

Fig. 3.6. Ground- and excited-state absorption spectra of ruby ($L=3$ cm) normalized to 100% population

given here, because all grating experiments, up to now, have been performed with a detection wavelength outside this wavelength region.

The grating was detected by Bragg diffraction of a quasi-cw Ar laser beam with a wavelength of 514 nm. The difference of the absorption coefficients $K_0 - K_1$ is rather small in this wavelength region. Therefore the diffraction efficiency is expected to be mainly determined by the phase grating contribution [3.31]. The grating decay time was found [3.32] to agree with the life-time of the ^2E states of about 3 ms for low Cr^{3+} concentrations of 0.05%. At higher Cr^{3+} concentrations, the grating decay becomes nonexponential and much faster than the fluorescence decay [3.33]. This indicates that, at higher Cr concentrations, not only the population of the ^2E levels has to be considered in the grating formation but also other levels, e.g., Cr–Cr pair levels.

The population-density grating in ruby is accompanied by a thermal grating, if the ^2E levels are excited via some higher absorption band, as indicated in Fig. 3.5. The excess energy is then converted into heat. The thermal grating has a shorter decay time than the population density grating so that both contributions are easily distinguished [3.33].

After the first experiment [3.18] transient gratings in ruby have been investigated to detect diffusion (see also Sect. 5.6) of the excited-state energy [3.32–35]. These experiments made it possible to give upper limits to the diffusion constant in agreement with spectroscopic investigations [3.36, 37].

It was suggested to use ruby for real-time holography [3.31, 38]. However, since the diffraction efficiencies are rather low (e. g., 3 % for a 2.3 cm long crystal with a pump power of about 700 W/cm^2 at 514 nm wavelength) other materials are currently used for holographic applications (see also Sect. 3.6 and Chap. 6).

3.4.3 Gratings in Various Crystals and Glasses

In addition to ruby, laser-induced transient gratings have been investigated in other solids. Experiments in the organic crystal p-terphenyl doped with pentacene were performed to observe energy transport between the excited states of pentacene [3.39]. The grating technique (four-wave mixing) was also used to investigate excited state (exciton) diffusion in the laser material $Nd_xLa_{1-x}P_5O_{14}$ [3.40]. It can be expected that the method will be applied to other materials (Sect. 5.6).

In Nd:YAG, phase conjugation has been demonstrated, which can be understood in terms of diffraction at a population density grating [3.41]. In principle, any laser material can be used for this purpose since gain saturation is inherent to all lasers leading to spatial hole burning (Sect. 7.1) and thus grating build-up.

Alkali halides are another class of materials where laser-induced gratings are readily produced [3.42–44]. In the most simple case, photo-excited free electrons are trapped at positive ion vacancies and produce color centers (F-centers). Optical transitions in the centers change the absorption of the initially transparent crystal. The mechanism of grating formation is more complicated than in the case of a simple population density grating. The color-centers and the corresponding gratings are usually stable. Although the emphasis in this book is on transient gratings, the color-center gratings are mentioned here because the distinction between permanent and transient gratings is somewhat arbitrary. Color-centers, for instance, may become unstable when an additional light beam is used to read out the grating. The absorption of this laser beam by the color-centers may lead to their ionization and to erasure of the grating while it is being read. In addition, transient effects occur during the build-up of color centers, like electron diffusion and configurational changes [3.45].

Grating experiments were also performed with amorphous solids, in particular with filter glass [3.47, 49, 52]. The glass samples investigated were used as saturable absorbers to Q-switch solid-state lasers.

3.4.4 Gratings in Dye Solutions

Laser-induced gratings in dye solutions were observed by a number of independent authors [3.46–48]. Different types of dyes can be employed. The first dyes investigated were saturable absorbers (laser Q-switches) and other dyes without fluorescence and with fast decay times, during which the absorbed light energy is rapidly converted into heat. Later, laser dyes with strong fluorescence were also used.

After the initial demonstration of the grating effect, it was shown that the optical properties of the dye solutions are changed in a twofold way. First, the incident intensity modulates spatially the population of the electronic states of the dye molecules and the absorption is bleached. This population density grating affects the amplitude of a probing beam and was originally called an amplitude grating. According to the Kramers-Kronig relation (3.30) also a phase contribution exists. Secondly, a temperature grating is caused by the heat that is set free by relaxation of the excited molecules. Since the refractive index of the solvent varies with temperature, this thermal grating corresponds mainly to a phase grating (Sect. 3.7). Here only the population density grating is considered which is dominant using short pulse (picosecond) excitation.

If a dye solution is excited with polarized light, the resulting change in the optical properties is anisotropic, i.e. the solution becomes dichroic and birefringent. The reason for the anisotropy is that the dye molecules are excited preferentially if the transition moment $\boldsymbol{\mu}$ is parallel to the electric field $\boldsymbol{E}$ of the light [3.53]. The grating cannot be described any more by a simple spatially modulated population density. Instead, the orientational distribution $\varrho(y, \theta, \phi, t)$ of the excited dye molecules has to be considered, where the orientation is given by the polar coordinates θ and ϕ (not to be confused with the pump and diffraction angle in Chap. 2). The excitation rate calculated from semiclassical radiation theory is given by

$$(\partial \varrho / \partial t)_{\text{exc}} = C |\boldsymbol{E} \cdot \boldsymbol{\mu}|^2 N_a / 4\pi . \tag{3.40}$$

Here $\boldsymbol{E}$ is the total field strength. $N_a / 4\pi$ gives the isotropic orientational distribution of molecules in the ground-state. Weak excitation is assumed so that the total number of excited molecules

$$N_b = \iint \varrho \sin \theta \, d\theta \, d\phi \tag{3.41}$$

is much smaller than the ground-state density N_a.

The coupling constant C can be expressed in terms of the absorption cross-section σ by considering linear polarization for $\boldsymbol{E}$. In this case, the excitation rate according to (3.40) is proportional to $|E|^2 |\mu|^2 \cos^2 \theta$. Integration gives $\partial N_b / \partial t = C |E|^2 |\mu|^2 N_a / 3$. Introducing the intensity by $I = |E|^2 / 2Z$, Z being the wave impedance, and comparison with (3.36) for $N_b \ll N_a$ gives

$$C = 6 Z\sigma / |\mu|^2 h\nu . \tag{3.42}$$

For grating excitation, the electric field strength is given by superposition of two plane waves

$$E = A_a \exp(i\boldsymbol{k}_a \cdot \boldsymbol{r}) + A_b \exp(i\boldsymbol{k}_b \cdot \boldsymbol{r}) . \tag{3.43}$$

A_a and A_b are the electric field amplitudes. Interestingly, grating excitation is possible with perpendicular orientation of A_a and A_b, as will be discussed later. In the case of parallel polarization of A_a and A_b, the excitation rate is given by

$$|E \cdot \mu|^2 = (|A_a|^2 + |A_b|^2 + 2|A_a||A_b| \cos qx)|\mu|^2 \cos^2 \theta \ . \tag{3.44}$$

Here $qx = (k_1 - k_2) \cdot r$ and θ is the angle between the transition moment μ of a dye molecule and the field strength $A_{a,b}$.

If excitation is performed with a short pulse (pulse width $t_p \ll$ orientational relaxation time and excited singlet life-time τ_s) the spatial amplitude of the resulting orientational distribution of the excited molecules is given by

$$\varrho(\theta, t = 0) = \frac{3}{4\pi} \sigma \frac{E_p}{h\nu} N \cos^2 \theta \tag{3.45}$$

with $E_p = |A_a||A_b| t_p / Z$. If $|A_a| = |A_b|$ the total excitation energy density is equal to E_p. The further temporal development of $\varrho(\theta, t)$ is given by the rotational diffusion equation

$$\frac{\partial}{\partial t} \varrho = D\nabla^2 \varrho - \frac{1}{\tau_s} \varrho \ . \tag{3.46}$$

D is the rotational diffusion constant and ∇^2 is the Laplace-operator

$$\nabla^2 = \frac{1}{\sin \theta} \frac{\partial}{\partial \theta} \left(\sin \theta \frac{\partial}{\partial \theta} \right) + \frac{1}{\sin^2 \theta} \frac{\partial^2}{\partial \phi^2} \ . \tag{3.47}$$

The solution of (3.46) with the initial condition (3.45) is given by

$$\varrho(\theta, t) = \frac{1}{4\pi} \sigma N \frac{E_p}{h\nu} \exp(-t/\tau_s) [\exp(-t/\tau_{or}) (3 \cos^2 \theta - 1) + 1] \ . \tag{3.48}$$

The orientational decay time is defined by $\tau_{or} = 1/6 D$. For simple molecules $\tau_{or} = \eta V/kT$, where η is the viscosity of the liquid, and V the volume of the molecule [3.53]. The change of the susceptibility depends on the polarization of the probing radiation E_p according to $|\mu| \cdot |E_p|^2$. If the polarization is parallel to the excitation field, one obtains

$$\Delta\chi_{\parallel} = -3(\alpha_0 - \alpha_1) \iint \varrho(\theta, t) \cos^2 \theta \sin \theta \, d\theta \, d\phi$$

$$= -\sigma N \frac{E_p}{h\nu} (\alpha_0 - \alpha_1) \exp(-t/\tau_s) \left[1 + \frac{4}{5} \exp(-t/\tau_{or}) \right] \ . \tag{3.49}$$

It is assumed that the dipole moments for transitions from S_1 to higher levels have the same direction as for the $S_0 \to S_1$ transition. The factor 3 has been chosen so that the upper line in (3.49) reduces to (3.35) for an isotropic distribution $\varrho = \Delta N / 4\pi$.

For perpendicular polarization of the probing beam with respect to the excitation, one obtains

$$\Delta\chi_\perp = -3(\alpha_0 - \alpha_1) \iint \varrho(\theta, t) \sin^2\theta \sin^2\phi \, d\theta \, d\phi$$

$$= -\sigma N \frac{E_p}{h\nu} (\alpha_0 - \alpha_1) \exp(-t/\tau_s) \left[1 - \frac{2}{5} \exp(-t/\tau_{or}) \right]. \tag{3.50}$$

Equations (3.49, 50) show that the total susceptibility $\chi + \Delta\chi$ of the dye solution becomes anisotropic, i.e. $\Delta\chi$ has to be considered as a tensor. Because of symmetry, the principal axes of the $\Delta\chi$ tensor are parallel and perpendicular to the polarization direction of the excitation beam. For $t=0$ the ratio of the principal values is $\Delta\chi_\parallel / \Delta\chi_\perp = 3$. For $t \gg \tau_{or}$ the $\Delta\chi$ tensor becomes isotropic.

The time dependence of the change of the optical constants is given by the lifetime of the excited electronic state (S_1) of the dye and by orientational relaxation. For many laser dyes, the oritentational relaxation time τ_{or} (typically several 100 ps) is faster than the S_1 life-time (typically several ns). For a complete description of the time dependence of the optical constants, it is also necessary to consider the triplet population and the temperature effect which decay on a microsecond to millisecond time scale.

The grating method has been used by several authors [3.54–57] to measure the time constants of the various decay processes. Another direction of current research on laser-induced gratings in dye solutions is stimulated by possible applications in real-time holography and phase-conjugation [3.58, 59 and further references given in these papers, see also Chap. 6].

Quantitative results on laser-induced gratings will be discussed for the example of Rhodamine 6G solutions in the following.

3.4.5 Rhodamine 6G Solutions

Rhodamine 6G, like many other laser dyes, can be described by the level diagram of Fig. 3.7. The electronic levels are divided into a singlet manifold $S_0, S_1, S_2, \ldots$ and a triplet manifold $T_0, T_1, \ldots$ connected to the singlet manifold by means of transitions with low probability. Each of the electronic levels is very wide in energy, being broadened by a continuum of vibrational, rotational and solvent states. At room temperature, only the lowest vibrational-rotational states of an electronic level are populated.

By light absorption with cross-section $\sigma = \sigma_{a0}$, some excited state of S_1 is produced. The molecules then relax rapidly to the lowest vibrational level of the S_1 state in which the population density is accumulated during the singlet life-time τ_s. Some molecules make transitions into the triplet state T_0. For Rhodamine 6G, the quantum yield for the production of triplet molecules from excited singlet molecules is very low so that the population of T_0 can be neglected specially for pulsed excitation. Regarding the optical pumping cycle, the molecules behave like 3-level-systems, because only the S_0-ground-state, the

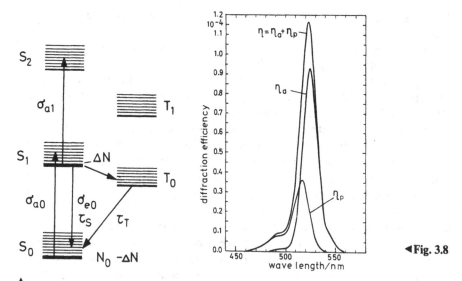

Fig. 3.7. Dye molecule energy level diagram

Fig. 3.8. Diffraction efficiency η consisting of an amplitude η_a and phase grating contribution η_p calculated for a 10^{-4} mol/l Rhodamine 6G solution in methanol with thickness $d=0.1$ mm. Energy density of excitation beams 10^{-4} J/cm². Unpolarized excitation assumed [3.61]. Excitation wavelength equals detection wavelength

S_1-state excited by the primary absorption process and the lowest vibrational S_1-state are of importance. For pulsed excitation (pulse width: $\ll \tau_s$) the change ΔN of the population density is determined by the incident energy density according to (3.39).

The change of the absorption coefficient ΔK due to optical excitation can be calculated from the cross-sections σ_{a0} for absorption from the S_0 ground state and σ_{a1}, σ_{e1} for absorption into higher states and stimulated emission back to S_0 from the lowest S_1 level

$$\Delta K = -(\sigma_{a0} - \sigma_{a1} + \sigma_{e1})\Delta N \ . \tag{3.51}$$

Experimental results for the cross-sections σ_{a0}, σ_{a1}, σ_{e1} are given in [3.60]. With these experimental values, ΔK can be calculated directly and the refractive index change Δn is obtained using the Kramers-Kronig relation (3.30). The resulting diffraction efficiency $\eta = \eta_a + \eta_p$ is given [3.61] in Fig. 3.8. The numerical values in Fig. 3.8 are only preliminary because the experimental values of the cross-sections in (3.51) are not yet finally established.

Two types of grating experiments will be described in the following. We consider first a self-diffraction experiment (Sect. 2.4.3) intended to determine the diffraction efficiency. A frequency-doubled Nd:YAG laser with 530 nm wavelength and mode-locked to produce pulses with 20 ps halfwidth has been used for grating excitation and detection. In a second experiment, the time evolution of the diffraction efficiency has been investigated using a probe pulse with variable delay.

In Fig. 3.9, the energy self-diffracted at a laser-induced grating in a Rhodamine solution is given as a function of incident energy density E_p. The grating constant amounted to $\Lambda = 20\,\mu m$. At low incident energy densities, the diffracted energy density varies approximately proportional to E_p^3. The diffraction efficiency is therefore given by $\eta \sim E_p^2$. This is expected, because for small E_p the excited state population density is linearly dependent, $\Delta N \sim E_p$. From (3.32, 35) it follows that the diffraction efficiency varies quadratically with the incident energy density $[\eta \sim (\Delta N)^2 \sim E_p^2]$, as observed in the experiment. At high incident energy density, a beginning saturation of the diffracted energy is observed which is explained by a saturation of the excited state population ΔN with incident energy density E_p according to (3.39). A complete calculation of the diffraction efficiencies from (3.32, 35, 39) using numerical values for ΔK and Δn starting from (3.51) gives only rough agreement with the measured efficiencies given in Fig. 3.9. This is probably caused by the limited accuracy of the experimental data used for the various cross sections to determine ΔK. Also geometrical factors present a difficulty in deriving reliable experimental values for the diffraction efficiency. [3.61].

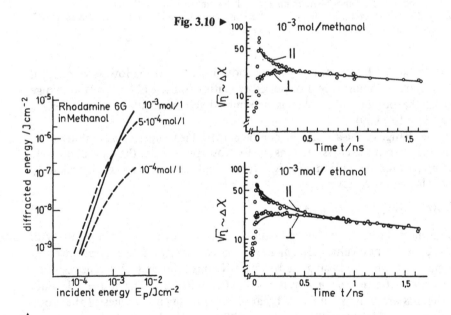

Fig. 3.9. Selfdiffracted vs incident energy density measured with different Rhodamine solutions (thickness 100 μm) and 530 nm laser pulses with 20 ps width

Fig. 3.10. Time dependence of the diffraction efficiency η of population density gratings in Rhodamine 6G in various solvents with probing beam polarization parallel ‖ and perpendicular ⊥ to excitation polarization η and $\Delta\chi$ are in arbitrary units

In Fig. 3.10, the time dependence of the diffraction efficiency of a grating in rhodamine solutions in methanol and ethanol is shown [3.55]. Grating excitation has been performed again with 20 ps pulses from a Nd:YAG laser. The grating has been probed with optically delayed pulses from the same source. The probe pulse polarization was parallel or perpendicular to the polarization of the excitation pulse. Because the diffraction efficiency η depends quadratically on the change of the optical constants or the susceptibility (3.22), the square root of η is given. The susceptibility change is clearly anisotropic.

Matching the calculated curves (3.49, 50) to the experimental values, the orientational relaxation time has been determined to be 100 ps for the dye in methanol and 170 ps in ethanol. These values are about 40 % smaller than the orientational relaxation times measured by the transient absorption method [3.53]. The difference is probably caused by lifetime shortening due to stimulated emission. The relaxation time τ_{or} in methanol is shorter than in ethanol because the higher viscosity of ethanol damps the orientational motion of the Rhodamine molecule and keeps it longer oriented. Also the singlet life-time $\tau_s \approx 2$ −3 ns has been determined and appears to be somewhat smaller than values known from other experiments [3.60].

Finally it should be noted in Fig. 3.10 that a sharp peak appears at $t = 0$. This coherence peak does not correspond to a rapidly relaxing change of the optical properties but is an artefact produced by interaction of the probing beam with one of the excitation beam as will be discussed in Sect. 7.7.

Further measurements on the diffraction efficiency of laser-induced gratings in dye solutions are necessary to give a complete quantitative description. The results will help to give a better understanding of the performance of dye lasers which are strongly influenced by grating formation (spatial hole burning), e.g., it has been suggested that these gratings are important for the formation and stabilization of femtosecond pulses in colliding pulse ring lasers [3.62].

3.4.6 Perpendicular Polarization of the Grating Excitation Beams

Grating excitation is also possible with perpendicular polarization of the excitation beams. This seems surprising since in this case no spatially modulated intensity develops. However, the orientational distribution of the excited molecules and also the optical properties exhibit a grating-like modulation. This can be understood by evaluating the excitation rate given by (3.40, 43) with $A_a \perp A_b$

$$(\partial\varrho/\partial t)_{exc} \sim |E \cdot \mu|^2 = |A_a|^2 |\mu|^2 \cos^2 \theta + |A_b|^2 |\mu|^2 \sin^2 \theta \cos^2 \phi$$
$$+ 2|A_a||A_b||\mu|^2 \cos \theta \sin \theta \cos \phi \cos qx . \tag{3.52}$$

The resulting spatial amplitude of the orientational distribution of the dye molecules is given in analogy to (3.48) by

$$\varrho(\theta, t) = \frac{3}{4\pi} \, \sigma N \frac{E_p}{h\nu} \exp(-t/\tau_s) \exp(-t/\tau_{or}) \cos\theta \sin\theta \cos\phi \ . \tag{3.53}$$

The maxima of the orientational distribution due to (3.53) appear at angles $\theta = 45°$ and $225°$. These directions are principal axes of the $\Delta\chi$-tensor. The principle values corresponding to these directions are

$$\Delta\chi_{45} \sim \iint (\cos\theta + \sin\theta \cos\phi)^2 \varrho(\theta, t) \sin\theta \, d\theta \, d\phi \ ,$$

$$\Delta\chi_{45} = -\Delta\chi_{225} \sim -\frac{3}{5} \, \sigma N \frac{E_p}{h\nu} \exp(-t/\tau_s) \exp(-t/\tau_{or}) \ . \tag{3.54}$$

The decay of $\Delta\chi_{45}$ is given by a single exponential with an effective decay time $(1/\tau_s + 1/\tau_{or})^{-1} \approx \tau_{or}$. This is a remarkable difference to the decay of $\chi_\parallel$ or $\chi_\perp$ which is given by the sum of two exponentials according to (3.49, 50). As it is easier to evaluate a decay time from a single exponential, it seems advantageous to measure τ_{or} in a grating experiment with perpendicular polarization of the excitation beams.

Grating excitation by perpendicularly polarized excitation beams has been observed in transient-transmission measurements [3.53] where a coherent coupling peak (Sect. 7.7) was found not only for parallel polarization of the bleaching and probing beam but also for perpendicular polarization.

3.5 Gratings in Semiconductors

In semiconductors a large number of grating investigations has been performed for the following reasons. First, forced light scattering at laser-induced gratings allows the study of changes in optical material properties due to a variety of nonlinear optical mechanisms. The absorption and refractive index changes are important for applications in real-time holography, phase conjugation, optical bistability, optical gating and laser mode-locking. These applications are facilitated because semiconductors are fairly well understood materials available in excellent quality and their physical properties can be controlled by chemical or radiation treatment. Second, laser-induced gratings in semiconductors are of interest in studying mobile electron and holes, especially their transport and decay parameters. These material properties are important for the fabrication and understanding of electronic and optoelectronic devices like transistors, photodetectors and lasers.

In this section, different mechanisms leading to light-induced changes of optical semiconductor properties are treated. Also, a review of materials is given where dynamic grating experiments have been performed. Investigations of grating decay processes due to transport and recombination of mobile carriers are discussed later (Sect. 5.5). Extensive reviews of nonlinear optical processes in semiconductors are available already in the literature, so that only a short summary is given referring to [3.63–65] for more details.

3.5.1 Mechanisms of Light-Induced Absorption and Index Changes

A shorthand compilation of the various effects leading to changes in the optical properties of semiconductors is given in Fig. 3.11. In the simplest case, the complex refractive index change $\Delta\tilde{n}$ is linearly dependent on the light intensity I, namely,

$$\Delta\tilde{n} = \tilde{n}_2 I .\tag{3.55}$$

The nonlinear refractive index $\tilde{n}_2$ is related to the often used third-order nonlinear susceptibility $\tilde{\chi}^{(3)}$ (in electrostatic units) by

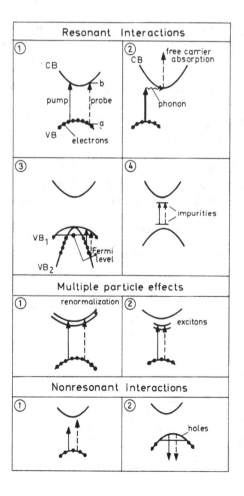

Fig. 3.11. Modulation mechanisms of the optical constants of semiconductors by light. (VB: valence band, CB: conduction band). The arrows indicate the energy of the pump and probe photons. These energies are equal in degenerate interactions.

Resonant interactions (real transitions, population changes): Single particle effects (polarizability changes): (1) Bleaching of the interband absorption (dynamic Burstein-Moss shift), (2) additional intraband absorption (sketched for indirect semiconductor Si), (3) intraband transitions (e.g., intervalenceband absorption in p-type semiconductors), and (4) transitions involving impurity levels.

Resonant interactions, Multiple particles effects: (1) Band-gap renormalization, and (2) saturation of exciton absorption, e.g., by carrier screening.

Nonresonant interactions (virtual transitions): (1) Intrinsic semiconductors: Nonlinearly bound carriers, and (2) doped semiconductors: free carriers in nonparabolic bands

$$\tilde{\chi}^{(3)}\,[\text{esu}] = 19\,n^2\tilde{n}_2\,[\text{cm}^2/\text{W}] \ . \tag{3.56}$$

In the case of resonant interactions, absorption of light results in a carrier density N given by

$$N = \zeta K \tau I / h\nu \ . \tag{3.57}$$

Here K is the absorption coefficient, τ the recombination time of an electron-hole pair and $h\nu$ the photon energy. The quantum efficiency $\zeta \le 1$ describes the fact that only part of the absorbed photons create electron-hole pairs. Absorption may also lead to heating or other material excitations. Often $\zeta = 1$ is assumed giving a crude estimate for N. Equation (3.57) is valid for cw excitation or long excitation pulses with a width $t_p \gg \tau$. If $t_p \ll \tau$, the recombination time τ in (3.57) has to be replaced by a time of the order of the pulse width [3.65]. The total index change $\Delta\tilde{n}$ is dependent on the carrier density

$$\Delta\tilde{n} = \tilde{n}_{\text{eh}} N \tag{3.58}$$

where the index change $\tilde{n}_{\text{eh}}$ per excited electron-hole pair is related to n_2 by

$$\tilde{n}_2 = \tilde{n}_{\text{eh}} K \tau / h\nu \ . \tag{3.59}$$

In general, $\tilde{n}_2$ and the nonlinear susceptibility are complex. However, refraction effects are often dominant and it is then sufficient to consider the real part of the complex quantities or the absolute values, e.g., $n_2 = \text{Re}\{\tilde{n}_2\} \approx |\tilde{n}_2|$. In Table 3.1, numerical values for n_{eh}, n_2 and $\chi^{(3)}$ are given for resonant interactions in selected materials.

For nonresonant interactions, the change of the optical properties is also described by (3.55, 56). Numerical values of nonresonant n_2 and $\chi^{(3)}$ have been given in [3.65]. Comparing resonant and nonresonant interactions, it can be

Table 3.1. Nonlinear refractive index n_2 and nonlinear susceptibility $\chi^{(3)}$ for selected semiconductors at wavelength λ and temperature T. Refractive index change per electron-hole pair: n_{eh}. Absorption coefficient: K. Carrier life-time: τ. Refractive index: n. The dominant mechanisms are discussed in the text

Material	T [K]	λ [μm]	n_{eh} [cm³]	K [cm⁻¹]	τ [ns]
Si	300	1.06	10^{-21}	10	100
InSb	77	5.03		50	
GaAs	< 100				
GaAs, MQW	300	0.848	5×10^{-20}	3600	20

Material	n_2 [cm²/W]	n	$\chi^{(3)}$ [esu]	Reference
Si	10^{-8}	3.56	10^{-6}	[3.66]
InSb	3×10^{-3}		1	[3.64]
GaAs	4×10^{-4}	3.4	10^{-1}	[3.64]
GaAs, MQW	2×10^{-5}		5×10^{-3}	[3.68]

stated that resonant changes of the refractive index are generally much larger than nonresonant effects. Resonant effects are due to population differences and their temporal response is thus limited to the range of $10^{-6}\ldots10^{-12}$ s. Nonresonant effects can be much faster perhaps with response times down to only some optical cycles.

3.5.2 Resonant Interactions, Single-Particle Effects

Absorption of light leads to real (not necessarily direct) transitions of the electrons so that excited states are populated. Changes of the optical properties can be produced by a couple of mechanisms which are classified as single- and multiple-particle effects.

Single-particle effects arise because the polarizability of the excited electrons is different from the ground-state polarizability. The situation is similar to that in Sect. 3.4 where similar polarizability changes are discussed, which are due to transitions of electrons localized to atoms or molecules. In semiconductors, the electronic transitions may take place from the valence to the conduction band (interband transition) or within a band (intraband).

a) Bleaching of the Interband Absorption and Resulting Refractive Index Changes

Interband transitions from the valence to the conduction band lead to a depletion of the absorbing electrons in the valence band. In addition, the density of unpopulated energy states in the conduction band is reduced. The absorption coefficient K is therefore bleached according to

$$K = K_0(f_a - f_b) \; , \tag{3.60}$$

where K_0 is the low intensity absorption coefficient and f_a, f_b are the population probabilities of the lower a and upper b states. Thermal population of the levels a and b can be neglected for sufficiently low temperature and doping. At zero incident intensity: $f_a = 1$ and $f_b = 0$. With increasing intensity, f_a decreases and f_b increases. For small excitation, as long as Boltzmann statistics is valid, the changes of f_a and f_b are proportional to the total excited carrier density N and therefore $\Delta K \sim N$. The absorption coefficient K is frequency dependent and can be used to calculate the refractive index in dependence of the optically induced population from the Kramers-Kronig relation. Absorption and index changes due to bleaching of the valence-conduction-band transition are important for direct band-gap semiconductors and have been investigated extensively for InSb, CdS and HgCdTe [3.65].

b) Additional Free-Carrier Absorption and Refraction

In indirect band-gap semiconductors, e.g., silicon, absorption of radiation (with a wavelength in the region of the band-gap, producing electrons in the conduction band) does not lead to absorption bleaching but, on the contrary, the absorption increases. This is due to intraband transitions in the conduction and valence bands. These transitions are not direct between allowed energy states but involve some other excitations like phonons. The bleaching of the interband absorption is small compared to the additional intraband or free carrier absorption.

The absorption and index changes due to the optically excited free carriers can be estimated with the help of the Drude model which treats the electron and holes classically as quasi-free carriers oscillating in the light field. The effective masses m_e and m_h of the electrons and holes reflect the band structure of the material. The induced optical dipole moment and polarizability of the electron and holes are easily calculated leading to the following expressions for the absorption and index changes [3.66]:

$$\Delta K = \frac{Ne^2}{nm_{eh}\omega^2 c\tau_d\varepsilon_0} = N\sigma_{eh} \ , \tag{3.61}$$

$$\Delta n = \frac{-Ne^2}{2\,nm_{eh}\omega^2\varepsilon_0} = Nn_{eh} \ . \tag{3.62}$$

Here N is the density of the optically excited electron hole pairs, $m_{eh} = (1/m_e + 1/m_h)^{-1}$ the reduced effective mass of an electron hole pair, and τ_d describes the damping of the carrier oscillation due to some scattering mechanisms; e is the charge of an electron or hole, ω the circular frequency of the light wave and of the resulting carrier oscillation, ε_0 the permittivity of vacuum and n the undisturbed refractive index of the material; n_{eh} gives the refractive index change for one electron-hole pair per volume element and σ_{eh} the absorption cross-section.

c) Bleaching of Intraband Absorption and Resulting Refractive Index Changes

Intraband transitions within the valence or conduction bands are possible if free carriers (electrons or holes) are present. These can be produced by optical excitation as discussed in the preceeding section or by doping. For example, holes are present in p-type materials and transitions between the sub-bands VB_1 and VB_2 (light and heavy hole band) are possible resulting in the so-called intervalence band absorption. The absorption and refractive index may change now due to incident pump intensity similarly as for interband transitions.

d) Transitions Involving Impurities

A further possibility to produce index and absorption changes via resonant interaction and utilizing single particle effects occurs in semiconductors with impurities. Experimental work with such materials has been scarce until now [3.65]. Absorption and index changes are expected to occur by saturating transitions between impurity levels. Related effects have been discussed already in Sect. 3.4 dealing with changes of the optical properties due to electronic transitions in atoms and molecules (population density gratings).

3.5.3 Resonant Interactions, Multiple-Particle Effects

a) Band-Gap Renormalization

High densities of free carriers change the band structure of a semiconductor because the carriers are able to rearrange in the electric field of the cores. This modifies the spatially periodic potential distribution and consequently the band structure. The band-gap shifts due to this renormalization and the optical properties change.

b) Saturation of Exciton Absorption

At low temperature, the optically excited electron-hole pairs from excitons, excitonic molecules and electron-hole liquids. Up to now mainly excitonic effects have been investigated experimentally for the production of light-induced changes of the optical properties. An exciton consists of an electron and a hole bound together by Coulomb attraction. This system has hydrogen atomlike energy states at energies E_n below the bottom of the conduction band

$$E_n = -E_b/n^2 \quad n = 1, 2, 3 \ldots . \tag{3.63}$$

The energy levels in the conduction band correspond to the ionization continuum of the hydrogen atom. The ionization energy is equivalent to the binding energy of the exciton $E_b = |E_1|$. Due to the high relative permittivity ε of semiconducting materials and the low reduced effective mass m_{eh} of an electron hole-pair compared to the free-electron mass m, the exciton binding energy E_b is much smaller than the hydrogen ionization energy of 13.6 eV:

$$E_b = (m_{eh}/m\varepsilon^2) \, 13.6 \, \text{eV} . \tag{3.64}$$

Typically E_b is in the range of 1 to 300 meV for different semiconductors [3.67]. The energy levels of the exciton are thus usually located slightly below the conduction band. These energy levels can be populated by direct optical transitions from the valence band. At low optical intensities, the absorption is linear. At higher excitation, the absorption bleaches which has been explained taking into account many-body effects [3.67]. At high exciton densities, where the interparticle spacing becomes comparable to the Bohr radius of the exciton, the Coulomb attraction between the electrons and hole becomes screened. The bound state becomes unstable. This screening ionization is called a Mott-transition. The excitons disappear and there is no more absorption. As usual, the absorption saturation is accompanied by a nonlinear refractive index change.

Excitonic effects are observed at low enough temperatures where the thermal energy of the particles is small compared to the binding energy. In multiple quantum well structures (MQW), the binding energy can be increased up to four times compared to the bulk material. This has been utilized in a GaAs/GaAlAs structure to observe optical absorption saturation at room temperature with optical powers as low as several milliwatts [3.68].

3.5.4 Nonresonant Interactions

a) Intrinsic Semiconductors

Light with a frequency below the bandgap does not produce transitions from the valence to the conduction band. In intrinsic (undoped) semiconductors, the valence band is filled completely whereas the conduction band is empty. Therefore no transition within the bands are possible. The electrons are only virtually excited to some energy level in the bandgap region. However, the bound electrons oscillate in the light field. Because of the complicated nonlinear restoring forces, the oscillation is anharmonic so that a nonlinear polarization is set up. In every material, a third-order nonlinear susceptibility $\chi^{(3)}$ exists which can be related to an intensity dependent complex refractive index.

b) Doped Semiconductors

In doped semiconductors, the electrons and holes may move or oscillate in the partially filled bands, as has been mentioned already in connection with the Drude model. This motion is anharmonic if the band is nonparabolic. Again, the result is a nonlinear susceptibility $\chi^{(3)}$ which is equivalent to an intensity dependent refractive index.

3.5.5 Material Investigations

a) Silicon

Gratings have been investigated first by *Woerdman* with regard to real-time holographic applications [3.69–71]. A Q-switched Nd:YAG laser was used for excitation and detection of the gratings. The energy of a laser photon $h\nu = 1.164$ eV is slightly larger than the optical energy gap $E_q = 1.112$ eV of Si at 295 K so that free carriers are produced by a single photon band-to-band absorption process. It was shown experimentally that the grating excited by spatially periodic illumination of the Si sample is a phase grating. The change of the refractive index can be calculated (Fig. 3.12) from the Drude formula (3.62) and was found to agree with the experiment. The grating decay time was measured and found to be given mainly by ambipolar diffusion

$$\tau_D = \Lambda^2/4\pi D_a \, , \tag{3.65}$$

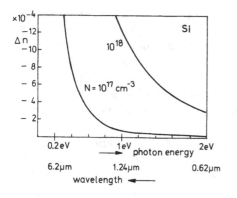

Fig. 3.12. Refractive index changes Δn in Silicon calculated from Eq. (3.62) with $n_{en} = 10^{-21}$ cm^3 at $\lambda = 1.06$ µm according to [3.66]. (N: free carrier concentration)

where Λ denotes the grating period and $D_a \approx 10$ cm^2/s the ambipolar diffusion constant. The decay times are typically in the range of 10 to 100 ns for grating periods Λ of about 10 to 30 µm. Later work of *Odulov* and co-workers [3.72] confirmed the original assumption that the grating corresponds to a periodic modulation of free carriers. The same group also performed experiments in crossed electric and magnetic fields which shift the free carrier grating in relation to the exciting interference pattern. This shift results in energy transfer between the two writing beams [3.73]. Gratings in Si were also investigated by *Jarasiunas* and co-workers [3.74, 75] who confirmed the previous results and showed that a high excitation also a thermal phase gratings is observable. This group used also a frequency-doubled Nd : laser of 6.53 µm to produce free-carrier gratings near the crystal surface which were probed at 1.06 µm. Such a configuration enables to investigate free carrier plasmas at high densities up to 10^{19} cm^{-3} in Si or even 10^{20} cm^{-3} in GaAs [3.187]. *Jain* and co-workers [3.76, 77] performed degenerate four-wave mixing experiments which can be understood also with help of the grating picture, and derived effective nonlinear susceptibilities $\chi^{(3)}$ for steady-state and pulsed excitation. At high excitation, saturation of the diffraction efficiency was observed and attributed to increasing free-carrier absorption.

All the experiments discussed so far have used the self-diffraction technique. In a three-beam experiment [3.78], grating excitation and detection was separated and detailed measurements of the dependence of the diffraction efficiency on the excited free-carrier density were possible. The Bessel-function dependence (Chap. 4) characteristic for a phase grating was observed. Shortening of the grating decay times due to Auger recombination was detected at high excitation levels. Further information on carrier dynamics is given in Sect. 5.5.

Phase conjugation in Si has been demonstrated by various authors [3.79, 80] investigating especially the quality and distortion of the conjugate beams.

b) Germanium

Volume gratings have been detected first accidentally in picosecond experiment where the time-dependent absorption of Ge excited with 5 ps pulses from a

Nd:glass laser has been probed with noncollinear delayed pulses from the same source [3.81–83]. Interference between the pump and probe beams leads to grating build-up, in addition to the absorption change, to be investigated. Diffraction at this grating produced an additional power change of the probe beam if the delay time between the pump and probe pulse is smaller than the coherence time of the pulses. This coherent coupling peak has been observed also in other pump and probe experiments with picosecond pulses and is discussed in more detail in Chap. 7.

Volume gratings in Ge have been investigated in detail in a series of papers by *Smirl* and co-workers [3.84–90]. The main aspects of their work are generation of a forward-traveling phase-conjugate wave [3.84], measurement of the diffusion coefficient and recombination effects [3.85] especially at high excitation [3.86], and observation of anisotropic filling of the electronic states in k-space [3.87–89]. The anisotropic state filling was observed first in a two-beam experiment with delayed excitation and probe pulses. A coherent coupling peak has been observed not only for parallel polarization of the two beams but also for perpendicular polarization where pump-beam interference does not produce a spatially modulated intensity distribution (Sect. 2.2). However, if the material has a polarization memory, a grating can be produced also with perpendicular polarization of the probe beams as has been discussed already in Sect. 3.4 (population density gratings). In semiconductors, a polarization memory is provided by anisotropic state filling. The randomization time of a distribution in the k-space to an isotropic distribution may be a short as 10^{-14} s.

Temporary *surface gratings* have been formed by interference of two beams with a wavelength of 694 nm from a ruby laser [3.91]. Theoretical and experimental work of *Wiggins, Herman* et al. [3.91, 92] indicated that the gratings are due to thermal reflectivity changes caused by heating of the surface by the nanosecond time duration pulses. The relatively long pulses and short wavelength result in heating rather than free-carrier generation. *Vaitkus* et al. [3.93] proposed that surface gratings on Ge, observed in reflected beam self-diffraction, are due to a thermal modulation of the crystal density, leading to a transient surface relief. The observed value of the diffraction efficiency was explained by assuming a relief amplitude of 10 nm.

Permanent diffraction gratings are observed at high pulse energies or powers [3.91]. Periodic surfaces structures (ripples) with a spatial frequency of approximately the laser wavelength can also be produced with a single laser beam [3.94]. The ripples arise from stimulated scattering of the incident laser light into surface waves at the air solid interface. The waves build up due to the optical excitation. These surface waves grow exponentially from spontaneous scattering via various possible feedback mechanisms. The final surface structure results from melting coupled with diffusional processes during the regrowth. Similar surface ripples have been observed also on other materials (see also Sect. 5.7).

c) III–V Compounds (Ga, In-P, As, Sb)

Galliumarsenide. The first grating experiment with GaAs was performed by *Hoffmann* et al. [3.95] using a picosecond Nd: glass laser at 0.53 μm. The surface recombination velocity was extracted from the decay of the diffraction efficiency after excitation. *Vaitkus* et al. [3.188] recorded volume gratings in GaAs by picosecond pulses and measured carrier lifetimes of 75 ps. Another grating experiment in a GaAs multiple quantum well structure (MQW) was undertaken by *Hegarty* et al. [3.86] with a tunable picosecond dye laser demonstrating that the diffraction efficiency (and thus the absorption and index changes) are strongly enhanced at the exciton resonance.

Further information on absorption and index changes due to injected free carriers in GaAs and related materials can be obtained from the following papers [3.97–102], [3.67, 68] covering theoretical and experimental work not using the grating method. Absorption and refractive index spectra calculated by *Haug* et al. [3.67] are given in Fig. 3.13.

Indium Phosphide. *Hoffmann*, *Jarasiunas* and co-workers [3.95] measured the surface recombination velocity also in InP with the grating method similar as in GaAs.

Indium Arsenide. A ruby laser with 20 ns pulse duration was used to asses InAs as a candidate for optical amplification by self-diffraction [3.103]. Like the previous experiments in Ge performed by *Wiggins* et al. [3.91, 92] and *Herman* et al. [3.93], the transient reflectivity grating was due to a change in the refractive index by surface heating.

Indium Antimonide. A large amount of experimental and theoretical work on nonlinear index and absorption changes in InSb has been performed by *Smith*

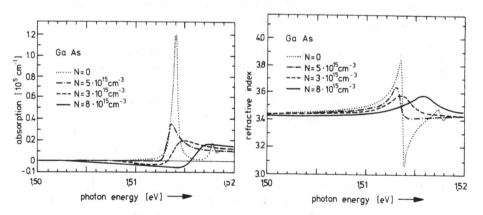

Fig. 3.13. Calculated absorption and refractive index spectra for GaAs with electronic temperature of 10 K for various carrier concentrations after *Haug* et al. [3.67]. The absorption maxima are due to exciton production, see Sect. 3.5.3b

and *Miller* using a CO lasers at about 5 µm wavelength. A good account of this work is given in [3.64]. Grating or degenerate four-wave mixing experiments [3.104] have been done to investigate carrier diffusion [3.189].

d) II–VI Compounds (Zn, Cd, Hg, − O, Se, S, Te)

Zinc Oxide. In ZnO, gratings have been produced with a ruby laser and detected by diffraction of a cw argon laser [3.105]. Because the band gap is much larger than the photon energy, mainly 2-photon absorption leads to optical excitation of the material. A 2-photon absorption process produces a change of the optical constants Δn, ΔK which is given by the square of the incident power density I

$$\Delta K, \quad \Delta n \sim I^2 \tag{3.66}$$

in contrast to a single-photon absorption process where

$$\Delta K, \quad \Delta n \sim I . \tag{3.67}$$

The diffraction efficiency in the case of two-photon absorption becomes $\sim I^4$ in contrast to single photon absorption with $\sim I^2$. The measured diffraction efficiency indicates a two-photon absorption process for grating excitation. A combined free-carrier and temperature grating has been assumed as grating mechanisms. Related theoretical work is described in [3.106].

Zinc Selenide. Experiments in ZnSe were performed to measure the ambipolar diffusion constant [3.107] and to demonstrate phase-conjugation by degenerate six-photon mixing [3.108]. Two-photon excitation using a ruby laser produced the grating by two-photon absorption. Diffraction at such grating is called six-photon mixing because 6 waves are involved; the two excitation waves are counted twice. A theory of six-photon mixing with a description by scattering diagrams has been developed [3.108].

(ZnSe)$_x$–(GaP)$_{1-x}$. Investigations of dynamic grating recording and erasing processes have been carried out in (ZnSe)$_x$–(GaP)$_{1-x}$ mixed crystals. Auger recombination limits the diffraction efficiency at high carrier concentrations. Auger recombination and free carrier absorption cause additional heating and result in the formation of thermal gratings [3.109].

Cadmium Sulfide. In CdS preliminary self-diffraction experiments with a ruby laser indicated that a spatially modulated free carrier or exciton distribution is the dominant grating mechanism [3.110]. Because the CdS crystals are transparent for the ruby wavelength the gratings were assumed to be excited by two photon transitions of electrons to the conduction band.

Experiments in CdS were also performed [3.111] with frequency-doubled Nd:YAG laser excitation. Although the bandgap $E_G = 2.42$ eV is larger than the photon energy $hv = 2.37\,eV$ a quadratic dependence of the diffraction efficiency on the energy density of the exciting Nd:YAG laser has been observed at low energy densities. A single-photon process is therefore assumed for the grating excitation which may be explained by absorption of impurity levels. Also, the fluorescence decay time measurements indicate that impurities contribute to the grating.

In CdS doped with Cu, gratings have been produced by 1-photon impurity-conduction band transitions for low intensities of the exciting ruby laser. At high intensities, 2-photon band-to-band transitions occured [3.112].

Diffraction at a free carrier grating in CdS was used to measure the ambipolar diffusion constant [3.107]. Tuning of the crystal temperature above 300 K enabled to optimize the phase-conjugation efficiency [3.190].

The biexciton in CdS has been studied by nonlinear coherent mixing [3.113]. The experiments can be interpreted with the grating picture, too. Additional extensive grating work in ZnO and CdS has been done recently by *Klingshirn* et al. [3.191, 192].

Cadmium Selenide. Free carrier gratings have produced with laser wavelengths of 1.06 and 0.694 μm. Because CdSe has a direct bandgap of 0.71 μm, carrier generation takes place by two-photon interband absorption or by excitation from impurity levels [3.114, 115]. The measured self-diffraction signal at 1.06 μm varies with I^4 which can be understood as a result of a mixture of these effects.

The energy of a ruby laser photon (1.786 eV) is close to the band gap (1.825 eV at 13 K) so that temperature tuning can be used to produce either free carriers or excitons. When the sample temperature was adjusted so that the photon energy plus that of an optical LO phonon (0.27 eV) was coincident with the energy of an $A(n=1)$ exciton, a large diffraction efficiency has been obtained indicating resonant interaction [3.116]. Further grating work in the vicinity of the exciton line showed free carrier screening effects at high excitations [3.193, 194].

If the sample temperature is not tuned for exciton excitation, free carriers are generated. It is then possible to measure the ambipolar diffusion constant [3.107].

Cadmium Telluride. Free carrier gratings in CdTe [3.117] and forward [3.65] and backward-direction four-wave mixing [3.117] have been studied. A Nd:YAG laser at 1.06 μm with a photon energy (1.17 eV) much less than the band gap (1.605 eV) was used. The generation of free carriers is therefore attributed to two-step absorption via intermediate impurity states [3.65]. Grating decay measurements gave carrier lifetimes of 600 ps [3.188].

Cd S_x Se$_{1-x}$ – Doped Glasses. These ternary semiconductor alloys are obtainable inexpensively in the form of sharp cut-off glass filters. Each filter glass consists of microcrystals with 100 to 1000 Å sizes with a fixed composition x suspended in a glass matrix. An interesting consequence for gratings in these materials is that grating decay by diffusion does not take place. The diffraction efficiency for sufficiently long pulses is therefore independent of the grating period [3.65].

$Hg_{1-x} Cd_x Te$. Very large third-order susceptibilities of up to $\chi^{(3)} = 10^3$ esu are expected for these materials with suitable choice of alloy composition ($x \approx 0.15$), temperature ($T \leq 4$ K) and wavelength ($\lambda \geq 100$ μm). With a CO_2 laser at 10.6 μm a third-order susceptibility of 5×10^{-2} esu has been measured which nearly has the same magnitude as the value obtained for InSb (Table 3.1).

The mechanisms for the optical nonlinearity are free carrier generation [3.118, 119] and conduction-band nonparabolicity for n-type samples [3.120].

Copper Chloride. Nonlinear spectroscopy by resonant coherent scattering was performed in CuCl to study the Γ_1 biexciton level in CuCl [3.121, 122]. The technique corresponds to self-diffraction at a laser-induced grating.

Using resonant excitation at 389 nm, excitonic molecules were excited in CuCl with a grating-like density modulation [3.123]. Excitons were produced after radiative relaxation. A probe beam with a broad spectrum (2 nm) was used to detected the gratings corresponding to the excitons and the excitonic molecules exploiting their resonant behaviour. The decay times were measured in dependence of the grating period and diffusion constants derived.

3.6 The Photorefractive Effect in Electrooptic Crystals

The photorefractive effect has been discovered originally as an undesirable optical damage in nonlinear and electro-optical crystals [3.124]. Light-induced changes of the refractive index have limited the usefulness of crystals with large electro-optic and nonlinear optical coefficients such as $LiNbO_3$, because the index changes gave rise to decollimation and scattering of laser beams in devices such as modulators and frequency doublers. Subsequently, materials exhibiting such an optical damage effect – later called the photorefractive effect – were applied as holographic recording materials [3.125].

The following basic model for the effect was proposed [3.126] and substantiated by further work reviewed in [3.127–131]. When light of a suitable wavelength is incident on a crystal, photoelectrons are generated which migrate in the lattice and are subsequently trapped at new sites. The resulting space charges give rise to an electric field strength distribution in the material which changes the refractive index via the electrooptic effect.

The photorefractive effect has been detected in a number of different crystals, the major ones used for holographic applications are $LiNbO_3$, $LiTaO_3$, $KTa_xNb_{1-x}O_3$ (KTN), $BaTiO_3$ [3.124–131], $Bi_{12}SiO_{20}$ (BSO) [3.132–136], $Bi_{12}GeO_{20}$ (BGO) and $KNbO_3$ [3.137–142]. The energy required to obtain an appreciable photorefractive effect in BSO and $KNbO_3$ is close to the photographic sensitivity of silver-halide layers. Holographic gratings are mostly recorded with visible cw lasers (Ar, Kr, He-Cd, He-Ne). The storage times of the recorded holograms or grating decay times range from milliseconds for $KNbO_3$ [3.138], hours for KTN to months and years for $LiNbO_3$ [3.127] if a fixing process is applied (Chap. 6). The holographic gratings can be erased by uniform illumination.

Using pulsed lasers with high energy densities, short writing and erasing times can be obtained. Experiments with frequency-doubled Nd:YAG lasers at 530 nm showed that optical writing, reading and erasing of refractive index gratings are possible on a nanosecond time scale or even faster [3.146, 147].

3.6.1 Production of Electric Field Grating

The photorefractive effect is caused by free electrons which are released from donors (e.g., Fe^{2+} impurities in oxygen-octahedra ferroelectrics like $LiNbO_3$) by the incident light. For observation of this effect in a homogeneous material, it is necessary to have some spatial modulation of the light intensity. Periodic, grating-like excitation is particularly well suited for experimental observation of the photorefractive effect and also for its mathematical description. The spatial modulation of the light-intensity gives rise to a corresponding modulation of the electron and ionized donor densities (e.g., Fe^{3+} in $LiNbO_3$). Initially the negative and positive electrical charges of the electrons and ionized donors compensate so that there is no net space-charge. However, the electrons move by diffusion, under the action of an external field or due to the photovoltaic effect. The electrons are subsequently trapped by empty donors. Because of the movement of the electrons, there is a spatial difference in the excitation rate of the ionized donors and the trapping rate of the electrons resulting in a spatially modulated electrical charge. An electric field builds up which modulates the refractive index via the electrooptic effect.

The photorefractive effect is explained, in most papers, by electron movement. It has been shown, however, that holes must also to be considered in materials with a large initial concentration of traps [3.149–152].

The movement of the photoexcited free carriers can be effected by three different mechanisms: diffusion, drift (when an electrical field is externally applied) and the photovoltaic effect, as will be discussed in the following.

Diffusion. The development of the space-charge field E_{sc} is sketched in Fig. 3.14. The light intensity I excites ionized donors and electrons. In addition, a spatially homogeneous background electron density may be present. The electrons diffuse so that the spatial amplitude of the electron density is reduced when compared to the spatial amplitude of the ionized donor density. This amplitude difference gives rise to a space-charge distribution modulated in phase with the light

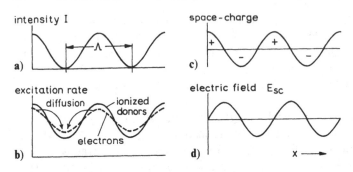

Fig. 3.14a-d. Build-up of a space-charge field E_{sc} by diffusion

intensity. The resulting electric field distribution E_{sc} is shifted by a quarter grating period $\Lambda/4$ against the light intensity.

For the further development of the electron density distribution, the influence of the space-charge field has to be considered in addition to diffusion. If diffusion dominates, the stationary electron density modulation is smoothed out completely [3.130]. On the other hand, if the space-charge field dominates the electron movement, the field prevents further smoothening of the electron distribution. In photorefractive crystals, the latter case is assumed and used for mathematical modeling [3.130, 131].

Drift. Displacement of the electron distribution can also be achieved by a static electric field. If the ionized donor excitation rate is proportional to $\cos qx$, the drifted electron excitation rate is proportional to $\cos(qx + \varphi)$ (Figs. 3.15 and 16). For sufficiently small displacements the difference of these distributions, i.e. the space charge density is proportional to $\sin qx$. The resulting electric field is proportional to $-\cos qx$ which corresponds to the intensity distribution except for the negative sign. An alternate way to understand the buildup of a spatially modulated field starts from the current produced by the applied dc voltage. In the stationary state, the current density J is constant as indicated in Fig. 3.15. The conductivity of the material is spatially modulated due to the modulation of the free carrier density. The electric field E_{sc} becomes therefore also spatially modulated.

Drift and diffusion are sufficient to explain the photorefractive effect in paraelectric crystals as KTN, BSO, BGO and in highly photoconducting ferroelectrics like $KNbO_3$ where the photovoltaic currents to be discussed below can be neglected.

Photovoltaic Effect. In piezoelectric crystals like $LiNbO_3$ a photocurrent can be produced also without an applied voltage. Photoelectrons are excited into the charge transfer band with a preferential direction of the velocity along the

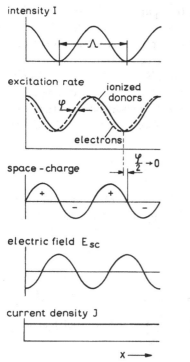

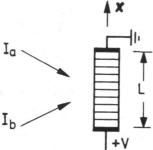

Fig. 3.15. Build-up of a space-charge field E_{sc} by an external dc voltage (Fig. 3.16) or by the photovoltaic effect. (φ: shift of the excitation rates, J: steady state current density due to applied voltage

Fig. 3.16. Photorefractive crystal with two writing beams and external dc voltage

Fig. 3.15   **Fig. 3.16**

direction of the polar axis. Additional current contributions due to anisotropic electron trapping and ion displacement are also possible. In total, the photovoltaic current density J_i^{ph} is given by

$$J_i^{ph} = -\beta_{ijk}E_j E_K^* \tag{3.68}$$

where E_j, E_K are electric field strengths components of the light wave and β_{ijk} the third-rank photovoltaic tensor which obeys $\beta_{ijk} = \beta_{ikj}^*$ and which has nonzero components only in media lacking a center of symmetry [3.154, 162]. Gratings in electrooptic crystals which are formed by photovoltaic charge transport mostly make use of the tensor component β_{333}. For this configuration

$$J^{ph} = -\beta_{333}E_3 E_3^* = -\beta_{333}I = -\kappa KI , \tag{3.69}$$

where I is the light intensity, the K absorption coefficient for light polarized along the x_3 direction and κ a constant characteristic of the crystal and doping.

The relation $J^{ph} = -\kappa KI$ is only valid for short electron transport length [3.130] where the electron transport length is defined as the mean distance the electrons travel before randomization of the velocity. The generalization to an arbitrary electron transport length will not be discussed here.

The photovoltaic effect produces a shift in the spatial distributions of the electrons and the ionized donors in a similar way as an applied dc field (Fig. 3.15). The electric field grating therefore is in phase with the excitation intensity distribution, except for the unimportant factor -1.

It has been shown in [3.162] that gratings can also be recorded by using nondiagonal components of the photovoltaic tensor. In these cases spatially oscillating photovoltaic currents can be produced by perpendicular polarization (ordinary and extraordinary) of the excitation beams (compare also Sect. 2.2.3). For such a light wave with

$$E = e_o E_o \exp(ik_o r) + e_e E_e \exp(ik_e r) \tag{3.70}$$

(e_o and e_e are the polarization unit vectors, E_o and E_e the complex amplitudes of the ordinary and extraordinary waves) an oscillating current appears in the crystal, e.g., along the x-direction if $\beta_{113} \neq 0$:

$$
\begin{aligned}
J_1^{ph} &= -\beta_{113}^s E_o E_e \cos(q \cdot r) - \beta_{113}^a E_o E_e \sin(q \cdot r) \\
&= \sqrt{(\beta_{113}^s)^2 + (\beta_{113}^a)^2}\, E_o E_e \cos[q \cdot r - \arctan(\beta_{113}^s / \beta_{113}^a)] \;.
\end{aligned}
\tag{3.71}
$$

This current consists of two parts, which are attributed to the symmetrical and antisymmetrical components of the photovoltaic tensor.

3.6.2 Basic Equations

The most general set of material equations describing the photorefractive effect has been given by *Kukhtarev* et al. [3.131]:

$$\frac{\partial n_e}{\partial t} = \frac{\partial N_D^+}{\partial t} + \frac{1}{e}\,\nabla J \;, \tag{3.72}$$

$$\frac{\partial N_D^+}{\partial t} = \left(\frac{\sigma}{h\nu}\,I + \beta\right)(N_D - N_D^+) - \gamma_R n N_D^+ \;, \tag{3.73}$$

$$J = e\mu n E + eD\nabla n + J^{ph} \;, \tag{3.74}$$

$$\nabla(\varepsilon\varepsilon_0 E) = e(N_D^+ - N_A - n) \;. \tag{3.75}$$

Equation (3.72) describes the difference of the electron and ionized donor excitation rates, as sketched in Figs. 3.14 and 15. The difference is due to the electronic current density J. The excitation rate of the ionized donors is given by (3.73), where $\sigma = \zeta K/N_D$ is the cross-section for photo-ionization which is related to the total absorption coefficient K via the quantum efficiency ζ. Thermal excitation is described by the rate constant β which produces a spatially homogeneous electron background, the donor density is N_D, the ionized donor density N_D^+, the electron density n_e, the recombination constant γ_R. The intensity I is given by

$$I(x) = I_0(1 + m \cos qx) \ , \tag{3.76}$$

where $I_0 = I_a + I_b$ is the total intensity of the two waves producing the grating, and $m = 2\sqrt{I_a I_b}/(I_a + I_b)$ the modulation ratio.

The current density due to the electric field E, due to diffusion and the photovoltaic effect (3.71) is given by (3.74), where e is the electron charge, μ the mobility and D the diffusion constant. The electric field is given by the external voltage applied along the crystal length L (Fig. 3.16) and the photoinduced space-charge field E_{sc}

$$E = E_{sc} + \frac{V}{L} \ . \tag{3.77}$$

Equation (3.75) gives the connection between the space-charge field E_{sc} and the charge density. ε is the relative permittivity of the medium. N_A is the number of ionized donors (empty traps) which are present without illumination. These are necessary to trap the excess number of electrons that arrive due to diffusion or drift from high-to-low-intensity regions. To have charge neutrality without illumination an equal number N_A of compensative acceptor levels has to be present.

A slightly simplified form of the material equations (3.72–75) has been given by *Moharam* et al. [3.130]:

$$J(x, t) = eD \frac{\partial n_e(x, t)}{\partial x} + e\mu n_e(x, t) \left[E_{sc}(x, t) + \frac{V}{L} \right] + J^{ph} \ , \tag{3.78}$$

$$\frac{\partial n_e(x, t)}{\partial t} = g(x) - \frac{n_e(x, t) - n_D}{\tau} + \frac{1}{e} \frac{\partial J(x, t)}{\partial x} \ , \tag{3.79}$$

$$E_{sc}(x, t) = -\frac{1}{\varepsilon \varepsilon_0} \int_0^t J(x, t) \, dt + G(t) \ , \tag{3.80}$$

where $G(t)$ has to be determined from

$$\int_0^L E_{sc}(x, t) \, dx = 0 \ . \tag{3.81}$$

Equation (3.78) corresponds to (3.74) assuming that all the quantities vary only in the x-direction. Equation (3.79) is obtained by combination of (3.72 and 73) with the approximation of constant donor density $N_D \gg N_D^+$. The free-carrier life-time is given by $(1/\tau) = \gamma_R N_D^+ \approx \gamma_R N_A = \text{const}$. The free-carrier concentration in the dark is $n_D = \beta N_D \tau$. The generation rate is given by $g(x) = g_0 I(x)/I_0$, where $g_0 = \sigma N_D I_0/h\nu = K\zeta I_0/h\nu$.

Equation (3.80) can be obtained from (3.75) using the continuity equation between space-charge and current

$$e \frac{\partial}{\partial t} (N_D^+ - N_A - n_e) + \nabla J = 0 . \tag{3.82}$$

3.6.3 Steady-State Space-Charge Field

In steady state $\partial N_D^+/\partial t = 0$, $\partial n_e/\partial t = 0$ and also $\partial J/\partial x = 0$ according to (3.82). Therefore from (3.79), we get

$$n_e(x) = n_D + \tau g(x) = (n_D + \tau g_0)(1 + m_1 \cos qx) \tag{3.83}$$

with

$$m_1 = \frac{\tau g_0}{n_D + \tau g_0} m . \tag{3.84}$$

Because $\partial J/\partial x = 0$ and $J = \text{const}$, the space-charge field can be directly calculated from (3.78). The current J is eliminated using (3.81) thus obtaining

$$E_{sc} = \frac{m_1 E_D \sin qx}{(1 + m_1 \cos qx)} + \left(E_V - \frac{V}{L}\right) \left(1 - \frac{\sqrt{1 - m_1^2}}{1 + m_1 \cos qx}\right) , \tag{3.85}$$

$$E_D = \frac{qD}{\mu} = \frac{qkT}{e} , \qquad E_V = \frac{\kappa h\nu}{e\mu\tau\zeta} \tag{3.86}$$

(k: Boltzmann constant, T: absolute temperature, κ: photovoltaic constant, Eq. (3.69)). The Einstein relation $D = \mu kT/e$ between mobility μ and diffusion constant D has been used.

The space-charge field given by (3.85) is nonsinusoidal and contains higher spatial harmonics. Fourier decomposition yields

$$E_{sc} = -2 E_e \sum_{h=1}^{\infty} \left[\left(\frac{1}{m_1^2} - 1\right)^{1/2} - \frac{1}{m_1}\right]^h \cos(hqx - \varphi_e) , \tag{3.87}$$

$$E_e = \sqrt{E_D^2 + (E_V - V/L)^2} , \qquad \varphi_e = \arctan(E_D/E_V - V/L) . \tag{3.88}$$

For material investigations, it may be desirable to suppress the higher spatial harmonics. This can be done by background illumination so that m_1 gets small. On the other hand, for other applications a large value of the grating field is obtained for $m_1 = 1$. The fundamental component of E_{sc} is given for maximum contrast $m_1 = 1$, i.e. $I_a = I_b$, $n_D = 0$, by

$$E_{sc}^1 = 2 E_D \sin qx - 2 (E_V - V/L) \cos qx . \tag{3.89}$$

Equations (3.85, 87–89) show that the space-charge field E_{sc} can be decomposed into three contributions with amplitudes of the order of $m_1 E_D$, $m_1 E_V$, $m_1 V/L$ originating from the diffusion, photovoltaic or drift mechanism. The amplitude of the saturated space-charge field due to diffusion is given by $E_D = q \cdot 0.026$ V at room temperature. A grating period $\Lambda = 2\pi/q = 1\,\mu\text{m}$ corresponds to $E_D = 1600$ V/cm. Smaller Λ give larger fields and vice versa. External fields V/L of up to 20 kV/cm are applied. The photovoltaic effect in ferroelectric crystals gives equivalent fields E_V up to 100 kV/cm.

According to (3.85–89) the saturated space-charge field does not depend on the light intensity. It will be shown in the following that the intensity determines the buildup time of the field.

Trap-Density Limited Space-Charge Fields. In the above analysis of the steady-state space-charge field it was assumed, that the trap density was sufficiently high to allow trapping at all photoexcited charge carriers. If the trap density is substantially lower, the photoinduced space-charge fields will be limited by the field produced by the trap charge density N_A [3.128]

$$E_q = \frac{eN_A \cdot \Lambda}{2\pi\varepsilon\varepsilon_0} \tag{3.90}$$

rather than the fields $E_V - V/L$ and $E_D = qkT/e$.

The influence of trap filling on the photorefractive effect can be described by introducing the space-charge screening length

$$l_D = \left(\frac{4\pi^2\varepsilon\varepsilon_0 kT}{e^2 N_A}\right)^{1/2} \tag{3.91}$$

and the length of electron tightening by the external electric field $E_0 = V/L$ [3.128]

$$l_E = \frac{2\pi\varepsilon\varepsilon_0 E_0}{eN_A} . \tag{3.92}$$

The photoinduced fields can then be expressed by these parameters and the maximum possible field E_q given by the trap density N_A

$$E_D = \left(\frac{l_D}{\Lambda}\right)^2 E_q \tag{3.93}$$

for the diffusion field, and

$$E_d = \frac{l_E}{\Lambda} \cdot E_q \tag{3.94}$$

for the drift field.

3.6.4 Short-Time Limit

The build-up of the carrier density $n_e(x, t)$ is given by (3.79). The carrier recombination time appears to be short, in the order of 10^{-9}–10^{-12} [3.6.4] in LiNbO$_3$:Fe, KNbO$_3$ and BaTiO$_3$, and comparatively long (≈ 1 μs) in Bi$_{12}$SiO$_{20}$ [3.167]. In the first group of materials therefore $n_e(x, t)$ follows directly (i.e., $\partial n_e / \partial t \approx 0$) the generation rate $g(x)$ and the current density $J(x, t)$ which change on a much slower time-scale under usual experimental conditions. Equation (3.78) can be simplified by neglecting the space-charge field E_{sc} which is small at the beginning. In addition, it has been assumed already, that the transport length of the photoelectrons is small compared to the grating period so that also J_{ph} can be neglected in (3.78). Solving (3.78 and 3.79) subject to these approximations gives

$$n_e(x) = n_D + \tau g_0 [1 + m' \cos(qx + \varphi)] \tag{3.95}$$

with

$$m' = m / \sqrt{[1 + (qL_D)^2]^2 + (qL_E)^2} \tag{3.96}$$

$$\varphi = \arctan \{ qL_E / [1 + (qL_D)^2] \} \tag{3.97}$$

where $L_D = \sqrt{\tau D}$, $L_E = \mu \tau V / L$ are the transport lengths associated with diffusion and drift. The current density J is obtained from (3.78) inserting (3.95, 69). From the current density the space-charge field is calculated using (3.80) which yields

$$E_{sc}(x, t) = \frac{e g_0 t}{\varepsilon \varepsilon_0} [m L_{ph} \cos qx - m' L_E \cos(qx + \varphi)$$

$$+ m' q L_D^2 \sin(qx + \varphi)] . \tag{3.98}$$

Here $L_{ph} = \kappa h \nu / e \zeta \ll 1$ is the transport length associated with the bulk photovoltaic effect. Equation (3.98) is composed of the contributions of the three mechanisms (photovoltaic effect, drift in external field and diffusion) producing the space-charge field. The relative magnitudes of the three contributions depend on the grating period $\Lambda = 2\pi / q$ and on the different transport lengths L_{ph}, L_E, L_D of the materials.

In LiNbO$_3$ all three processes are important. The transport lengths are much smaller than the usual fringe spacing Λ. In nonferroelectric crystals like BSO only drift and diffusion have to be considered. The transport lengths in BSO and reduced KNbO$_3$ are comparable to usual fringe spacings so that much smaller energy densities compared to LiNbO$_3$ are necessary to produce an equivalent photorefractive effect.

It should be noted that (3.98) is only valid for short times, i.e. $t \ll T_0'$ where T_0' is a time-constant describing the build-up of the space-charge field. For $t \gg T_0'$ the space-charge field E_{sc} reaches a time independent saturation value (Fig. 3.17)

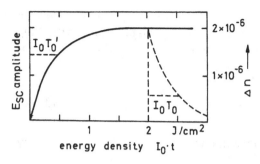

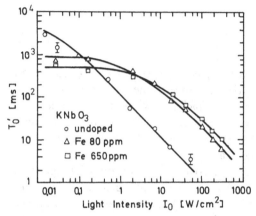

Fig. 3.17. Time dependence of the space-charge-field amplitude and corresponding refractive index change measured in a $LiTaO_3:Fe$ crystal with excitation wavelength $\lambda = 350{,}7$ nm after *Orlowski* et al. [3.152]

Fig. 3.18. Intensity dependence of the rise time T_0' for $KNbO_3:Fe$ [3.139]

given by (3.85). The calculation of the space-charge field for intermediate writing times has been done numerically [3.130] but will not be given here. As an approximation the rise-time is assumed to have the same form as the grating decay-time T_0 given by (3.101)

$$T_0' \approx \frac{\varepsilon\varepsilon_0}{e\mu(n_D + I_0 K\zeta\tau/h\nu)} \; . \tag{3.99}$$

In Fig. 3.18 as an example the experimental intensity dependence of the rise time T_0' as a function of light intensity is shown for different $KNbO_3:Fe$ crystals. At low light intensities T_0' is constant and given by the dark carrier concentration n_D of the crystals. As the writing intensity I_0 increases, the photoexcited carriers dominate, leading to T_0' inversely proportional to the absorbed light intensities. This behaviour is qualitatively in correspondence with (3.99). To obtain a quantitative understanding of the decay times possibly also drift and diffusion have to be considered as indicated in [3.138].

3.6.5 Grating Decay and Erasure

The photoelectrons are released from the traps by thermal excitation or by uniform illumination and move in the space-charge field to regions of high ionized donor concentration where recombination takes place. The electron density $n_e = n_D + \tau g_0$ is approximately constant during this process. The resulting decay of the space charge $e(N_D^+ - N_A - n)$ can be calculated from (3.72, 74, 75) which yield

$$\frac{\partial(N_D^+ - N_A - n_e)}{\partial t} = -\frac{1}{e} \nabla J = -\mu n_e \nabla E = \frac{-e\mu n}{\varepsilon\varepsilon_0} (N_D^+ - N_A - n_e) . \qquad (3.100)$$

The decay of the space-charge is exponential with a characteristic time given by the dielectric recombination time T_0

$$T_0 = \frac{\varepsilon\varepsilon_0}{e\mu n_e} = \frac{\varepsilon\varepsilon_0}{e\mu(n_D + \tau g_0)} = \frac{\varepsilon\varepsilon_0}{e\mu(n_D + I_0 K\zeta\tau/h\nu)} . \qquad (3.101)$$

A more rigorous treatment of grating writing and erase times where, in addition to the dielectric relaxation time, also the influence of the mean field drift time $T_E = (q\mu E_0)^{-1}$ and the diffusion time $T_D = e/\mu k T q^2$ have been considered, was discussed in [3.167]. In the regimes where the last two time constants become dominant the grating time constant can depend also on the applied electric field E_0 and the spatial frequency q of the grating.

The numerical values for T_0 and T_0' cover many orders of magnitude depending on the light intensity I_0 and the material parameters, mobility μ, absorption coefficient K, quantum efficiency for carrier generation ζ and carrier life-time τ. With high power pulsed lasers, T_0' can be in the order of 10^{-8} s, using cw lasers in highly photoconductive materials ($\mu\tau$ large) like BSO, BGO and KNbO$_3$ the rise and decay times are of the order of 10^{-2} s, and even larger in low conductivity LiNbO$_3$.

3.6.6 Electrooptic Refractive Index Changes

The space-charge field E_{sc} changes the refractive indices due to the electrooptic effect. For the simple case of a sinusoidal electric field pattern $E_{sc} = \Delta E \cos qx$, e.g., see (3.85) with $m_1 \ll 1$, the refractive index is modulated according to

$$n = n_0 + R E_{sc} = n_0 + \Delta n \cos qx \qquad (3.102)$$

with an amplitude

$$\Delta n = R \Delta E ,$$

$$= (n_3{}^3 r_{33}/2) \Delta E \quad \text{for} \quad \text{LiNbO}_3, \text{KNbO}_3,$$

$$= n^3 r_{41} \Delta E \qquad \text{for} \quad \text{BSO} , \qquad (3.103)$$

where R depends on the electro-optic tensor r_{ij} (Table 3.2) and the orientation of

Table 3.2. Some properties of photorefractive materials (K: absorption constant measured at wavelength λ, κ: photovoltaic constant, E_v: photovoltaic field, $\zeta\mu\tau$ characterizes the photoconductivity)

Material	λ [nm]	K [cm^{-1}]	κ [nAcm/W]	$\zeta\mu\tau$ [cm^2/V]	E_v [V/cm]	ε
LiNbO$_3$: 0.2% Fe	514.5	3.8	3.0	7×10^{-14}	10^5	30
LiNbO$_3$: 0.03% Fe	440	2.0	2.5	5.2×10^{-13}	10^4	
LiNbO$_3$	440	0.12	2.7	2×10^{-12}	5×10^3	
LiNbO$_3$ reduced	440	1.6	2.6	9×10^{-11}	8×10^1	
LiTaO$_3$: 0.02% Fe	488	1	2.3	10^{-13}	5×10^4	43
KNbO$_3$: 0.06% Fe	488	4.7	0.25	5×10^{-12}	1.5×10^2	50
KNbO$_3$	488	0.9	12	1.7×10^{-8}	1.8	
KNbO$_3$ red.	488	4.0	19	1.9×10^{-8}	2.6	
KNbO$_3$ red.	488	3.8	3	3×10^{-8}	2	
Bi$_{12}$SiO$_{20}$	514	6		1.4×10^{-7}		56

Material	Refractive index	Electrooptic constant [cm/V]	References
LiNbO$_3$: 0.2% Fe	$n_3 = 2.27$	$r_{33} = 31 \times 10^{-10}$	[3.141]
LiNbO$_3$: 0.03% Fe	$n_1 = 2.19$		[3.129]
LiNbO$_3$	at 700 nm		[3.158]
LiNbO$_3$ red.	$n_3 = 2.19$		
LiTaO$_3$: 0.02% Fe	$n_1 = 2.18$ at 600 nm	$r_{33} = 33 \times 10^{-10}$	[3.140]
KNbO$_3$: 0.06% Fe	$n_3 = 2.28$	$r_{33} = 64 \times 10^{-10}$	[3.141]
KNbO$_3$	$n_2 = 2.33$		[3.143]
KNbO$_3$ red.	$n_1 = 2.17$		[3.139]
KNbO$_3$ red.	at 633 nm		[3.161]
Bi$_{12}$SiO$_{20}$	$n = 2.53$ at 633 nm	$r_{41} = 3.5 \times 10^{-10}$	[3.155] [3.156] [3.157]

the crystal. The examples given for LiNbO$_3$, KNbO$_3$ and BSO refer to optimum crystal orientations.

For materials exhibiting no linear electro-optical effect, the refractive index changes due to the corresponding quadratic effect

$$n = n_0 + (R'/2) (E_a + \Delta E \cos qx)^2 \ . \tag{3.104}$$

If $E_a = 0$, the refractive index modulation has double spatial frequency compared to the space-charge field. The refractive index is modulated with the same spatial frequency q as the space-charge field if an additional bias field E_a is applied. The amplitude Δn of the fundamental component of the refractive index modulation

$$\Delta n = R' E_a \Delta E \tag{3.105}$$

can be altered with the bias field.

Numerical values for the refractive index change are of the order of up to $\Delta n \approx 10^{-3}$ for both linear and quadratic materials [3.127].

3.6.7 Multiphoton Photorefractive Effect

Photorefractive gratings in electrooptic crystals can also be recorded by multiphoton excitation. The principle of multiphoton photorefraction is shown in Fig. 3.19. The crystal is transparent at frequencies ω_1 and ω_2, yet by simultaneous absorption of one photon at ω_1 and another photon at ω_2 free carriers can be excited to the conduction band. The grating can be recorded at ω_1 (or ω_2) and only during recording the formation process is sensitized by the second frequency ω_2 (or ω_1). Reconstruction of the grating by light of frequency ω_1 does not erase the grating because no free carriers are generated due to the absence of the second frequency.

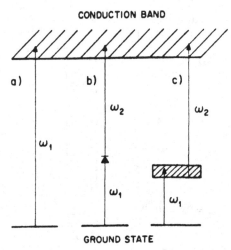

Fig. 3.19

Fig. 3.19. One and two-photon excitation mechanisms for photorefractive recording (a) Single-photon absorption: read-in, read-out and erasure with the same frequency ω_1, (b) two-photon absorption via a virtual intermediate level; read-in with frequencies ω_1 and ω_2, read-out at ω_1 (or ω_2), and (c) two-step absorption via a real intermediate level: read-in and erasure with ω_1 and ω_2, read-out at ω_2

Fig. 3.20. (a) A full two photon record-erase cycle. The refractive index amplitude Δn^{TP} is plotted versus the product of green and IR-intensities and time $I_{0.53} \cdot I_{1.06} \cdot t$. (b) The same experiment as in (a), but now Δn^{TP} is plotted versus the product of time and IR-intensity only. So it is possible to show that without a green beam ($I_{0.53} = 0$) no erasure occurs [3.163]

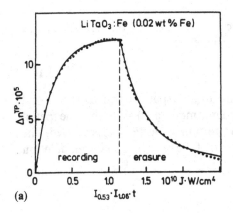

(a)

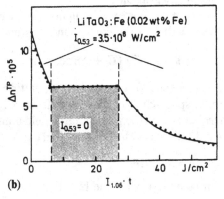

(b)

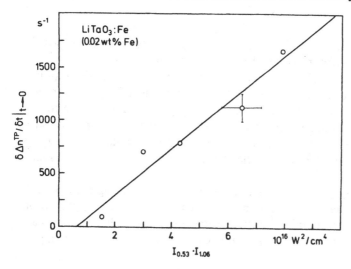

Fig. 3.21. Change of refractive index amplitude $\delta\Delta n^{TP}/\delta t|_{t\to 0}$ versus product of green and IR intensities $I_{0.53}\cdot I_{1.06}$. The slope of the solid line is the two-photon sensitivity S^{TP} [3.163]

This can be illustrated by the two-photon recording experiments in LiTaO$_3$:Fe performed by *Vorman* and *Krätzig* [3.163]. They recorded gratings via the intermediate ^5E-Fe^{2+}-state excited by the 1.06 µm emission line of a Nd:YAG laser (Fig. 3.20a). Simultaneous irradiation with the doubled laser-frequency (0.53 µm) caused electron transitions from the ^5E-state to the conduction band. The grating may be erased by a homogeneous illumination at both frequencies simultaneously. Reading the grating by the recording wavelength ($\lambda = 1.06$ µm) does not erase it (Fig. 3.20b).

For two-photon photovoltaic grating recording with pulse duration t_p much smaller than the excited state life time, the refractive index change at the beginning of the recording process ($t\to 0$) for small drift length and modulation ratio is

$$\Delta n^{TP}(t) = \frac{n^3 r \kappa K_{1.06}\cdot\sigma^* t_p}{\varepsilon\varepsilon_0 h\nu_{1.06}}\cdot I_{1.06}\cdot I_{0.53}\cdot t \ , \tag{3.106}$$

where σ^* is the excited state absorption cross section [3.163] and t the recording time of the mode-locked laser pulse train. In Fig. 3.21 it is illustrated that $(\Delta n^{TP}/t)_{t\to 0}$ depends linearly on the product at the intensities of the two recording intensities, as expected for two-photon recording.

The photosensitivity for two-photon recording has been shown to be much larger than for conventional one-photon processes in several materials including K(Nb, Ta)O$_3$ [3.164] and undoped LiNbO$_3$ [3.165], especially if long-lived real intermediate states such as Cr^{3+} in LiNbO$_3$ and LiTaO$_3$ have been used [3.166].

3.6.8 Applications

Much of the early interest in photorefractive materials came from possible applications in high volume optical storage. Research was directed into the development of both read-write and read-only systems and optimization of relevant material properties like storage and erasure sensitivity, nondestructive reading, storage time, storage volume, dynamic range, resolution, scattering etc. Table 3.2 gives a slight indication how the fundamental material properties can be changed by doping and other treatments. Different setups for hologram storage have been invented and special materials processing techniques were developed.

In recent years, photorefractive materials are increasingly applied in holography and image-processing, both in real-time. Image amplification, phase-conjugation, holographic interferometry and contour holography have been demonstrated. These and other applications are described in Chaps. 6 and 7.

3.7 Thermal Gratings

We call a grating thermal if the material excitation producing the optical grating is in local thermal equilibrium. Gratings in temperature and concentration fall into this class. Gratings in the population of very low-lying levels, such as the so-called two-level states in glasses or doped crystals, can also be of importance since they usually are coupled to the local thermal bath. As all of these fluctuations contribute to the entropy of the sample, they could also be called entropy gratings.

3.7.1 Temperature Gratings

Temperature gratings are the most common and almost inevitable reaction of a sample to irradiation by an interference pattern.

The dynamics of temperature gratings is governed by the equation of heat diffusion supplemented by a driving term which accounts for the absorbed pump radiation:

$$\frac{\partial T}{\partial t} = \nabla \cdot [D_{th}(T)\,\nabla T] + KI(r,t)/\varrho c \ . \tag{3.107}$$

If the thermal diffusivity is temperature independent one obtains

$$\frac{\partial T}{\partial t} = D_{th}\nabla^2 T + KI(r,t)/\varrho c \tag{3.108}$$

where T is the temperature and ∇ the nabla operator. *The thermal diffusivity* $D_{th} = \lambda/\varrho c_p$ is given by the heat conductivity λ, the density ϱ, and the specific heat

c_p at constant pressure. The temperature depencence of D_{th} can usually be neglected but needs consideration close to a phase transition and at very low temperatures. Further, λ and D_{th} are second-rank tensors in general. They degenerate to scalars in cubic crystals and disordered media which allows for the simpler form of (3.108). In the following discussion we shall use the latter equation, for simplicity. K is the absorption coefficient at the frequency of the pump beams, and $I(r, t)$ is their intensity. In the treatment of thermal gratings, we shall assume, for simplicity, that the two pump beams have equal polarization. The intensity is then given by (2.17), rewritten as

$$I = I_{av} + 2\Delta I \cos qx \ , \tag{3.109}$$

where $I_{av} = I_A + I_B$ and $\Delta I = (I_A I_B)^{1/2}$, compare (2.23). We further assume that the plane grating conditions (2.30–32) are fulfilled, i.e., that the pump beam interaction region is large compared to the grating wavelength and that the attenuation during passage through the sample is negligible. Under these circumstances, the temperature response to the absorbed pump radiation can be conveniently split into a slowly varying average change T_{av} and a grating structure $\Delta T \cos qx$. The two contributions are almost uncoupled under most experimental conditions. It is therefore possible to split (3.108) into two parts which can be solved independently

$$\partial \Delta T / \partial t + D_{th} q^2 \Delta T = 2 K\Delta I / \varrho c \ , \tag{3.110}$$

$$\partial T_{av} / \partial t - D_{th} \nabla^2 T = K I_{av} / \varrho c \ . \tag{3.111}$$

There are two relevant times associated with (3.110, 111), namely

$$\tau_q \equiv 1/D_{th} q^2 \ , \tag{3.112}$$

$$\tau_w \equiv w^2 / 8 D_{th} \ , \tag{3.113}$$

where w is the beam waist (see Fig. 2.1).

We first discuss the grating behavior. The solution to (3.110) is simple when the pump beams are rectangular pulses, short spikes or periodically modulated.

1) Rectangular pulses with duration t_p:

$$\Delta T = \Delta T_{st}(1 - e^{-t/\tau_q}) \ , \quad t < t_p \quad \text{where} \tag{3.114}$$

$$\Delta T_{st} = 2 K\Delta I / \lambda q^2 \tag{3.115}$$

is the steady-state value of the thermal grating amplitude. After the end of the pump pulse, the grating amplitude decays exponentially with the same time constant as before

$$\Delta T = \Delta T(t_p) e^{-(t - t_p)/\tau_q} \ , \quad t \geq t_p \ . \tag{3.116}$$

Thus, by recording the temporal dependence of the scattering, the thermal diffusivity can be determined immediately, sensitively, and in a contact-free manner.

2) The response to a short pump pulse with duration $t_p \ll \tau$ is

$$\Delta T_p = (2\,K\Delta I t_p / \varrho c)\, e^{-t/\tau_q} \; . \tag{3.117}$$

3) Amplitude modulation of the pump with frequency $\Omega / 2\pi$ results in a square-root of a Lorentzian for ΔT:

$$|\Delta T(\Omega)| = \Delta T_{st}(1 + \Omega^2 \tau_q^2)^{-1/2} \; . \tag{3.118}$$

Note that (3.118) when squared is identical to the respective constitutive equation for the classical thermal light scattering spectrum.

We now turn to the discussion of the average temperature rise. The steady-state solution to (3.111) strongly depends on boundary conditions, for example, on whether a sample is immersed into a temperature bath and on what is its geometrical shape; therefore, a general solution cannot be presented. The response to a pulse with duration $t_p \ll \tau_w$, on the other hand, can easily be calculated and helps understanding the total temperature development. For such a short pulse, the heat dissipation by diffusion out of the pumped zone is negligible while the pump pulse is on. Integration of (3.111) thus simply yields

$$T_{av}(x, t) = [KI_{av}(x)/\varrho c] \cdot t \; , \qquad t \le t_p \; . \tag{3.119}$$

TEM$_{00}$ laser beams are assumed for grating excitation so that $T_{av}(r, t) = T_{av}(x, t)$ has cylindrical symmetry around the z-axis.

After the end of the pump pulse, a different approach can be chosen for the solution of (3.111), provided the sample is sufficiently thick, namely $d/w \gg 1$. At time t_p, the sample is prepared with a Gaussian average temperature distribution with cylindrical symmetry. The decay of such a temperature distribution can easily be related to that of a pulsed line source along the z-axis which creates a cylindrical delta distribution at time $t = 0$. Following, for instance, [3.168] we can write for $t > 0$

$$T(x, t) \propto (1/4\pi D_{th} t) \exp(-x^2/4\,D_{th} t) \; . \tag{3.120}$$

This is a Gaussian distribution. The width w_{th} defined as, see (2.4),

$$w_{th}^2 / 2 = 4\,D_{th} t \tag{3.121}$$

increases with time, while the peak height decreases inversely.

$$T(0, t) \propto 1/4\pi D_{th} t \; . \tag{3.122}$$

The temperature distribution (3.120) can obviously be matched to that at the end of the pump pulses if the time origin is adjusted properly. The condition $w_{th} = w$ is satisfied at time τ_w, justifying our definition (3.113). Note that the width of the temperature distribution is that of the pump intensity, the width of which is $w/\sqrt{2}$.

When the time zero is assumed at the beginning of the pump pulses, the resulting expression for the average temperature distribution is

$$T_{av}(x, t) = [T_0(\tau_w/t + \tau_w - t_p)]$$

$$\cdot \exp[(-2x/w)^2 (\tau_w/t + \tau_w - t_p)^2] \quad \text{for} \quad t > t_p .\tag{3.123}$$

In Figs. 3.22 and 23, the total temperature distribution has been plotted for the condition $qw = 20\pi$. The time is measured in units of τ, i.e., $t' \equiv t/\tau_q$ and $\tau_w' = \tau_w/\tau_q = q^2 w^2/8 = 50\pi^2$. The value $qw = 20\pi$ was selected for clarity of presentation in the two figures; in most experiments, much larger values, say $qw > 100$ are used. The interference pattern has been assumed to be fully modulated, i.e., $I_A = I_B = \Delta I$, and normalized to yield $\Delta T_{st} = 2$. The pulse duration was chosen as $t_p' = 3$.

Figure 3.22a shows the temperature distribution for selected times during and immediately after the pulse. It is seen that the temperature grating is fully modulated at the beginning but, upon getting closer to its steady-state value, it begins to ride on an increasingly larger background of average temperature rise. The background is produced by the heat flow from the grating peaks to the valleys, filling them up more and more. After the pulse, the grating decays exponentially; it has practically disappeared at $t' = 6$, while the average temperature still is very close to its value at the end of the pulse, $t' = 3$. The time dependence of peak, valley and average temperatures at the center are shown in Fig. 3.22b.

Figure 3.23 shows the decay of the average temperature rise at selected times after the pulse. Note the drastically increased time steps as compared to Fig. 3.22. The slow decay shows that the pulse repetition rate has to be kept quite small if accumulation of pump beam heating is to be avoided.

The average temperature decay was derived under the assumption that the sample is thick. Frequently, $d/w \leq 1$ under experimental conditions. Then, heat is transported away not only radially but also in the z-direction through the ends of the sample. The decay then becomes faster, in particular if the neighboring medium has a large thermal diffusivity.

The overall decay, although slow, can be disturbing in a transient grating experiment if the probe beam is not well focused to the center of the interaction zone. The problem is that some probe light gets deflected from the slopes of the overall distribution which may blur the real signal. The simplest way to eliminate this problem is phase reversal in heterodyne detection which leaves the disturbance unaffected while the real signal changes sign. Thus, effective discrimination can easily be achieved [3.169, 170].

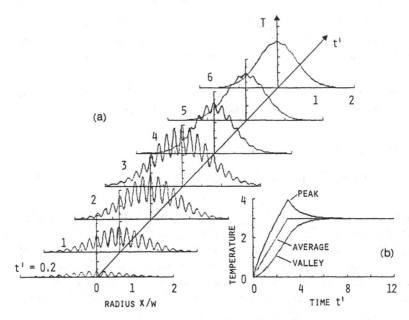

Fig. 3.22a,b. Temperature development during and immediately after a pump pulse forming an interference grating. (**a**) Temperature distributions in grating direction at selected, normalized times; (**b**) peak, valley, and average temperatures at the center

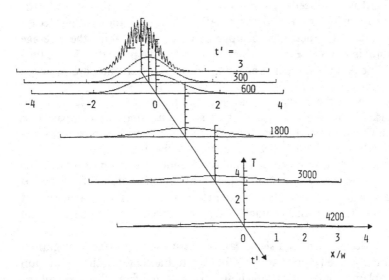

Fig. 3.23. Temperature development after the pump pulse

We now turn to the discussion of a rare but interesting phenomenon called second sound. Under very special conditions, heat can propagate through a medium in a wave-like rather than a diffusive manner. The best-known example for the existence of this so-called second sound is superfluid helium. Temperature and concentration of superfluid (vs normal fluid) liquid He couple in such a way that a restoring force arises. As a result, the heat-transport equation has to be supplemented by a second time derivative of the temperature converting [3.107] into a wave equation.

In crystals, thermal excitations can be represented by a gas of thermal phonons which behave in many aspects similar to the atoms of a real gas [3.171]. A temperature grating corresponds to a spatial variation in phonon density. In analogy to (first) sound in a real gas, one might indeed expect such a disturbance to propagate wavelike. The analogy, however, is a good one only when most phonons collide elastically. This is possible only at cryogenic temperatures and in very perfect crystals [3.171]. We wish to briefly discuss this rare phenomenon here, since one of the observations of second sound in a crystalline material was made by using FRS [3.172], see Sect. 5.2: As a result of the above-mentioned second derivative term in the heat-transport equation, the temperature response to pump modulation has a resonance at the second-sound frequency $\Omega_0 = q_0 c_0$ (c_0: speed of second sound) [3.171]

$$|\delta T(\Omega)| = \delta T_0 - \Gamma_R [(\Omega - \Omega_0)^2 + \Gamma_R^2]^{-1/2} \ . \tag{3.124}$$

Here Γ_R accounts for damping which is closely related to the probability of inelastic phonon scattering.

Finally, the case of strong absorption shall be briefly discussed. A fairly simple analytic solution to (3.108) exists for strong pump-light absorption and short pulse excitation. The product ansatz

$$T = T_\perp(x, t) \cdot g(z, t) \tag{3.125}$$

allows a separation which yields for $T_\perp$ the above-discussed plane grating solution. The explicit form of $g(z, t)$, a combination of exponentials and Erfc functions, represents the temperature response to absorption of an unmodulated plane wave. Since the diffracted probe light integrates over the sample thickness, in general, the exact shape of $g(z, t)$ is not very important in an FRS experiment with heavily absorbing samples.

3.7.2 Stress and Strain Gratings

Temperature gratings couple to strain because of thermal expansion of the medium. In fluids, these gratings are isobaric, but in solids expansion is restricted to directions along q [3.173, 174]. As a consequence, stress is created $\perp q$. In the simplest case, an amorphous substance or cubic crystal with $[100] \| q \| x$, the amplitudes of the stress $\vec{\sigma}$ and strain $\vec{u}$ gratings associated with δT are

$$\sigma_{yy} = \sigma_{zz} = \beta_{\text{eff}}(c_{12} - c_{11})\,\delta T \tag{3.126}$$

$$u_{xx} = \beta_{\text{eff}}\delta T \;, \tag{3.127}$$

while all the other tensor components vanish.

$$\beta_{\text{eff}} = \beta(1 + 2\,c_{12}/c_{11}) \tag{3.128}$$

is the effective or clamped thermal expansion coefficient appropriate for plane wave geometry, β the ordinary thermal expansion coefficient, and c_{11}, c_{12} are elastic constants [3.173].

3.7.3 Thermally-Induced Optical Gratings

The anisotropy expressed in (3.126) and (3.128) is carried on to the susceptibility grating $\Delta\chi$ associated with $\vec{u}$, and δT:

$$\Delta\chi_{ij} = (\delta\chi_{ij}/\partial T)_{\text{eff}}\,\partial T$$

$$= \delta_{ij}(\partial\chi/\partial T)_u\delta T - \varepsilon^2 p_{ijkl}u_{kl}(\partial T) \;. \tag{3.129}$$

The first term is isotropic and represents the variation of polarizability with phonon occupation (clamped conditions). *Wehner* and *Klein* [3.174] showed that this effect can be considerably larger than the second term in some substances. Particularly large values may be expected near a phase transition or in glasses because of structural rearrangement; and this does indeed seem to be the case [3.175].

The second term in (3.129) represents the influence of thermal expansion on the dielectric constant. p_{ijkl} are Pockel's elasto-optic constants, and ε the average dielectric constant. The microscopic origin of this term is to be seen in the variation of particle density at constant polarizability which yields an isotropic contribution to $\Delta\chi$, and the variation of polarizability α_{ij} with strain which is anisotropic:

$$p_{ijkl} = -(2/n^2)\,[\delta_{ij}\delta_{kl} + (\partial\alpha_{ij}/\alpha\,\partial u_{kl})] \;. \tag{3.130}$$

The last term in (3.130) – also connected with structural rearrangement – can be separated experimentally from the others by comparison of polarized and depolarized scattering.

3.8 Concentration Gratings

Concentration gratings are created either by chemical reaction [3.176] or by enrichment of one component in a mixture in the hot (cool) part of a thermal grating. The first method can be used conveniently, for instance, with photochromic materials; the second exploits the Soret effect to convert a temperature into a concentration grating [3.177]:

$$C = k_T \Delta T / T \ . \tag{3.131}$$

Equation (3.131) is valid for stationary conditions. C is the concentration and k_T is the Soret constant. The physical background to this relation is the minimum principle for the total free energy of a mixture which can be reduced by accumulating one component in the hotter zones of the sample. The quantity k_T is a constant of the order 1, but tends to diverge close to a critical consolute point. Conversion from temperature to concentration can then be fairly efficient.

The dynamics of a concentration grating is governed by the equation of mass diffusion. When thermally driven, it is to be supplemented by the driving Soret term

$$\partial C / \partial t = D_m [\nabla^2 C + (k_T / T) \nabla^2 T] \ . \tag{3.132}$$

Equation (3.132) needs a few comments:

1) For a complete account of the coupling between the thermodynamic variables C and T it is necessary to supplement (3.108) by a corresponding coupling term

$$\partial T / \partial t = D_{th} (\nabla^2 T + T k_T^{-1} \nabla^2 C) \ . \tag{3.133}$$

Thus a concentration grating, in principle, also creates a temperature grating (Dufour effect). The effect, however, is negligibly small in condensed media.

2) Heat and mass diffusivities typically are several orders of magnitude apart, for instance, $D_{th} = 0.1 \ cm^2/ms$ while $D_m = 1 \ cm^2/day$ for aqueous solutions. A primary thermal grating therefore decays after turning off the pump source, long before the concentration grating. The overall heating effect, on the other hand, may decay faster or be slower, depending on (qw) and D_{th}/D_m.

3) It is sufficient, therefore, to consider the thermal grating ΔT as a driving force only for the time the pump is on, and to neglect ΔT afterwards. The decay time then, simply, is

$$\tau_m = (D_m q^2)^{-1} \ . \tag{3.134}$$

Concentration gratings couple to light effectively since $\partial n / \partial C \simeq 0.1$. Hence, even very small amplitudes, $\Delta C / C < 1$ ppm, can easily be detected.

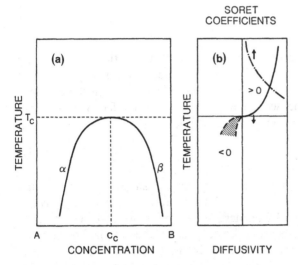

SORET
COEFFICIENTS

Fig. 3.24. (**a**) Phase diagram of a fluid mixture with a critical consolute point (c_c/T_c). (**b**) Soret coefficient and mass diffusivity near a critical point (schematic)

The study of concentration gratings is of particular interest near a critical consolute point of a mixture. Figure 3.24a schematically shows the phase diagram of such a mixture with components A, B in the concentration/temperature plane. The mixture is continuously miscible above T_c. Below T_c, a forbidden zone opens up where the mixture separates into two phases α, β of A- and B-rich material, respectively. Near T_c, the chemical potential which drives interdiffusion of the two species A, B becomes very flat (vs c); therefore, D_m is very small (Fig. 3.24b). The Soret coefficient, on the other hand, tends to diverge. Therefore, the coupling between thermal and concentration gratings is efficient near T_c; the decay, however, becomes extremely slow. At T_c, D_m is zero; below, at $T < T_c$, D_m is negative, resulting in a build-up of fluctuations or gratings which eventually lead to phase separation (shaded zone in Fig. 3.24). Thus, we encounter here a situation where a transient grating does not decay after turning off the excitation but tends to diverge! This effect has in fact been seen [3.178] (Sect. 5.2.2); it may help improve our understanding of phase separation processes in the future, which is of great technical importance in metallurgy and other areas of materials preparation.

3.9 Nonresonant Effects in Liquids

Gratings have been produced in transparent or *weakly absorbing liquids* using a high power Q-switched ruby laser for excitation and a cw argon probe laser. The principal contributions of the refractive index modulation arise from absorption, electrostriction, the electrocaloric effect and the Kerr effect. The main motivation for these experiments was to investigate the physical effects and gain mechanisms responsible for stimulated scattering [3.179–181].

Liquid crystals in the nematic phase can have very large nonlinear refractive indices. This is due to the reorientation of the director of the ordered phase by the optical field. Self-diffraction up to high orders is readily observable [3.172] using, e. g., a cw argon laser with a power of 2 W and a spot diameter of 1 mm. Experiments [3.183] suggest that also thermal gratings have to be considered.

Liquid crystal light valves are presently used for optical modulation of light beams. These light valves consist of a liquid crystal film subject to a perpendicular electric field. The electric field is spatially modulated by an optical field using a photoelectric electrode layer parallel to the film. The electric field, in turn, modulates the optical properties of the liquid crystal. Liquid crystal light valves thus provide an indirect opto-optical interaction. Very low energy densities of about 1 $\mu J/cm^2$ are required for operation. The erasure times of such liquid-crystal-photoconductor devices are comparatively slow, above 1 ms [1.4].

3.10 Related Phenomena

Laser-induced index and absorption changes in optical media have been investigated in the past not only in grating experiments but also extensively in connection with passive Q-*switching, mode locking of lasers, self-focussing, - defocussing* and *self-modulation of laser beams* [3.184].

Much work has been devoted to the *optical Kerr effect in liquids,* where a strong light field leads to an orientation of the molecules due to the interaction with a permanent or induced dipole moment. If the optical polarizability of the molecules is anisotropic, the medium becomes anisotropic and the refractive indices change. Additional effects are molecular rocking, libration and redistribution. Also electrostriction and the electrocaloric effect contribute to the intensity-dependent refractive index [3.184, 185].

More recently, the demonstration of optical *bistability* with Fabry-Perot resonators containing saturable absorbers or materials with an intensity-dependent index of refraction has attracted great interest because of possible applications in optical signal processing.

Experimental *data on refractive index changes* have been surveyed [3.186] for two classes of mechanisms: nonresonant effects (molecular-orientation Kerr and related effects, electrostriction and nonlinear electronic polarizability) and near-resonant effects which are associated with the saturation of an absorption line (Sect. 3.4).

4. Diffraction and Four-Wave Mixing Theory

Diffraction at permanent or dynamic gratings is a special problem in the more general context of light-matter interaction. The optical excitation of materials can be treated by different theoretical descriptions with varying levels of sophistication (Fig. 4.1).

The simplest approach is to consider the dependence of the complex refractive index on the incident-light intensity or energy. With this approach, a large number of laser-induced grating phenomena can be understood. Examples are given in the preceeding Chaps. 1–3. In the framework of nonlinear optics, diffraction at laser-induced gratings is described as optical mixing of the incident waves via a nonlinear polarization (Sect. 2.7). Even more fundamental is the use of the coupled Schrödinger and Maxwell equations where the spatial modulation of the optically excited states has to be considered. Due to this modulation, so-called spatial-coupling effects between the incident light-waves occur similar as in four-wave mixing and in laser-induced grating experiments. An example for an experiment where a quantum-mechanical description of matter is necessary is given in Sect. 3.5. In the theories mentioned so far, the light field is always treated classically. Aspects resulting from field quantization in four-wave mixing have been discussed in [4.1].

In the following we shall not try to outline all these theories but rather intend to give the derivation of simple formulas which are often used for the evaluation of experiments. The discussion starts with diffraction at permanent gratings and then switches to time-dependent effects.

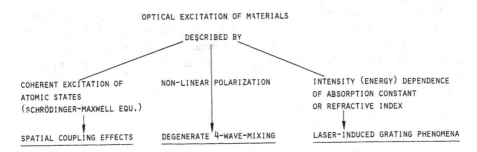

Fig. 4.1. Theoretical approaches to describe diffraction at laser induced dynamic gratings and related phenomena

4.1 Survey of Diffraction Theories

The literature on diffraction phenomena is abundant; thus since 1930 there have been over 400 scientific papers on permanent gratings [4.2]. We are mainly interested in volume gratings characterized by a spatial modulation of the absorption and refractive index. Surface gratings with a periodic thickness profile were discussed extensively in [4.3].

In the classical approaches of Huyghens and Kirchhoff, described in many textbooks, diffraction was treated as a boundary-value problem of the wave equation. The field strength of light incident on the grating or on another diffracting object is modulated in amplitude and phase. The resulting field strength just behind the grating gives the boundary conditions for the solution of the wave equation behind the grating.

The boundary-value method is appropriate for thin gratings. Examples for this method shall be given in Sect. 4.2. For gratings with arbitrary thickness [4.4], wave propagation inside the grating material has to be considered (Sect. 4.3) resulting in a set of coupled differential equations for the amplitude of the diffracted waves. Two limiting cases occur as characterized by the parameter $Q = 2\pi d\lambda/\Lambda^2 n$. For thin gratings with $Q \ll 1$, the amplitudes of the diffraction waves are given by the same Bessel-function expressions that are derived from the boundary-value method. For thick gratings with $Q \gg 1$, only one diffracted wave with large intensity is observed. This approximation is called the two-wave or Bragg case (Fig. 2.4).

The simplest possibility to describe diffraction at arbitrary thickness gratings is to consider planar gratings on which infinite plane waves are incident. The theories based on this assumption are called one-dimensional (1-D), because the amplitude of the diffracted waves produced in the gratings vary only with the coordinate perpendicular to the grating boundaries. In contrast, in 2-D theories, the amplitudes of the incident and diffracted waves may vary also along a direction parallel to the grating boundaries. 2-D theories [4.4, 5] are used to model the diffraction of Gaussian laser beams and also the behaviour of nonuniform gratings, which appear in holography. The following discussion emphasizes 1-D theories.

The most common methods of analyzing planar grating diffraction are the coupled-wave approach and the modal approach. In the coupled-wave approach, the total field in the grating is expressed as a superposition of plane waves (space harmonics) with amplitude varying in the direction perpendicular to the grating. These space harmonics inside the grating correspond to diffracted orders outside of the grating. Thus, the partial waves inside the grating medium are visualized as diffracted waves that progress through the grating slab and couple energy back and forth between each other.

Alternatively, in the modal representation, the field inside the grating is expressed as a superposition of the allowable modes (Bloch waves) of the periodic medium which are connected to the diffracted waves outside of the grating by appropriate boundary conditions. Both coupled-wave and modal approaches can produce exact descriptions of the diffraction problem without approximations. In their full rigorous forms, the formulations are completely equivalent [4.2]. In Sects. 4.3 and 4.4, examples for the application of the coupled-wave approach are given. The modal approach is not used in this volume.

After treating diffraction at permanent gratings in Sects. 4.2–4, the discussion in the following Sects. 4.5–7 switches to special effects which occur at light-induced dynamic gratings. In Sect. 4.5, various self-diffraction effects are treated where the diffraction of the beams inducing a grating is considered. Closely related are degenerate four-wave mixing experiments where diffraction of a signal beam at a dynamic grating produces a reflected beam which is phase-conjugated and amplified. Four-wave mixing may have important applications in pictorial information processing and is therefore discussed in the separate Sect. 4.6.

Finally, in Sect. 4.7, some diffraction properties of moving gratings are outlined. Such gratings are produced by interference of two light waves with different frequencies or with time-varying phase shifts between the two waves.

The description of dynamic gratings in Sects. 4.5–7 is based first on a generalization of the coupled-wave approach for permanent gratings. It is shown for some examples that a theory using a third-order nonlinear optical polarizability gives the same results.

4.2 Thin Transmission Gratings

A planar thin grating is described (Fig. 4.2) by a spatially periodic intensity transmittance $T(x) = T(x + \Lambda)$ or an amplitude transmittance $t(x)$ with $|t|^2 = T$. For real $t(x)$, an amplitude grating is obtained. If $t(x)$ acts only on the phase of a light wave penetrating the grating, this is called a phase grating. Generally $t(x)$ is complex and a mixed amplitude and phase grating is present.

An incident light wave is given by

$$E_i = \frac{A_i}{2} \exp\left[i(\omega t - kz)\right] + \text{c.c.} \ . \tag{4.1}$$

Normal incidence is assumed, for simplicity. The field strength just in front of the grating is given by $A_i \exp(i\omega t)$. The field strength just behind the grating is

$$E(z=0) = \frac{A_i}{2} t(x) \exp(i\omega t) + \text{c.c.} \ . \tag{4.2}$$

This field strength at the boundary $(z=0)$ radiates into the right half-space $(z>0)$. The field strength E for $z>0$ is described by a superposition of plane waves with the amplitudes A_m:

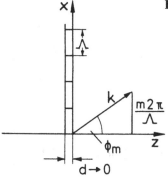

Fig. 4.2. Thin transmission grating geometry

$$E = \sum_m \frac{A_m}{2} \exp\left[i(\omega t - k_m x - \sqrt{k^2 - k_m^2}\, z)\right] + \text{c.c.} \tag{4.3}$$

where k_m is the x-component of the wave-vector of the mth partial wave. The z-component is chosen so that the absolute value of the wave-vector of a partial wave equals k. Waves with $k < k_m$ are evanescent in the z-direction. The amplitudes A_m are determined by matching (4.3) to the boundary condition (4.2), so that

$$\sum_m A_m \exp(-ik_m x) = A_i t(x) \tag{4.4}$$

Because $t(x)$ is periodic with Λ, (4.4) corresponds to a Fourier series development of $t(x)$ if k_m is chosen as

$$k_m = m2\pi/\Lambda , \qquad m = 0, \pm 1, \pm 2, \dots . \tag{4.5}$$

The directions of the diffracted waves are given by

$$\sin \phi_m = k_m/k = m\lambda/\Lambda . \tag{4.6}$$

It should be noted that this formula is valid for normal incidence only. The generalization to arbitrary incidence ($\alpha \neq 0$) is given by (2.42). The amplitude A_m of a diffracted partial wave is given by the mth Fourier coefficient of the transmittance $t(x)$:

$$A_m = (A_i/\Lambda) \int_0^\Lambda t(x) \exp(im2\pi x/\Lambda)\, dx . \tag{4.7}$$

4.2.1 Sinusoidal Amplitude Transmittance Grating

With the transmittance function

$$t(x) = \tau_0 + \tau_1 \cos(2\pi x/\Lambda) \tag{4.8}$$

one obtains

$$A_0 = \tau_0 A_i \; ,$$

$$A_{1,-1} = (\tau_1/2) A_i \; , \tag{4.9}$$

$$A_m = 0 \quad \text{for} \quad m \neq 0, \pm 1 \; .$$

A maximum diffraction efficiency of $\eta = (A_1/A_i)^2 = 6.25\%$ is obtained for 100% modulation of the transmittance, i.e. $\tau_0 = \tau_1 = 1/2$.

Information concerning the diffraction properties of nonsinusoidal transmittance gratings has been summarized in [4.6].

4.2.2 Absorption and Refractive Index Gratings

A spatial modulation of the complex refractive index $\tilde{n} = n + \Delta\tilde{n}\cos qx$ leads to an amplitude transmittance

$$t(x) = \exp\left[i\phi\cos(2\pi x/\Lambda)\right] \tag{4.10}$$

where

$$\phi = 2\pi\Delta\tilde{n}d/\lambda \; . \tag{4.11}$$

The amplitudes of the diffracted waves are given by

$$A_m = (A_i/\Lambda) \int_0^\Lambda \exp\left\{i\left[\phi\cos(2\pi x/\Lambda) + m2\pi x/\Lambda\right]\right\} dx \tag{4.12}$$

$$= (A_i/2\pi) \int_0^{2\pi} \exp\left(i\phi\cos t\right)\cos mt\, dt \tag{4.13}$$

$$= A_i i^m J_m(\phi) \; . \tag{4.14}$$

J_m is the mth order Bessel function defined by (4.13). For $|\phi| \ll 1$ the following approximations hold

$$J_0(\phi) \approx 1 \; ,$$

$$J_1(\phi) = J_{-1}(\phi) \approx \phi/2 \; . \tag{4.15}$$

With these approximations, the result (2.47) for the diffraction efficiency η is derived. Generally the diffraction efficiency of the first order is given by

$$\eta = |J_1(\phi)|^2 \; .$$

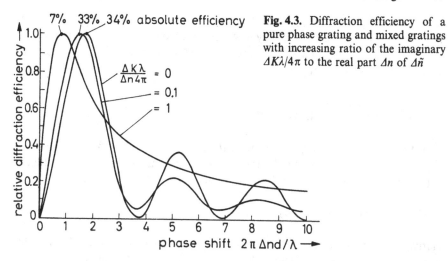

Fig. 4.3. Diffraction efficiency of a pure phase grating and mixed gratings with increasing ratio of the imaginary $\Delta K \lambda / 4\pi$ to the real part Δn of $\Delta \tilde{n}$

In Fig. 4.3, the diffraction efficiency is shown for various ratios of the real and imaginary part of $\Delta \tilde{n}$. A pure phase grating corresponds to real ϕ and $\Delta \tilde{n}$. In this case, a maximum diffraction efficiency of 34% is obtained for $\phi \approx 1.8$.

4.2.3 Limitations of the Theory

In the theory outlined above, the grating is described by a spatially modulated transmission function which modulates the incident plane wave. This implies that the optical properties vary only slowly in spatial dimensions comparable to the wavelength so that a plane wave approximation is possible. It is therefore expected that the outlined theory is valid only for small modulations of the complex refractive index. Diffraction at gratings with strong index modulations ($\Delta n \approx n$) seems to have received little attention up to now perhaps because such gratings have not yet been produced technically. Also, in transient grating experiments, the index modulation is usually small, and the theory outlined in this section is applicable.

To gain some understanding on the diffraction properties of gratings with strong index modulation, we quote some results on reflection gratings consisting of a corrugated surface with sinusoidal profile. A plane wave reflected at such a surface is phase-modulated like a wave transmitted by a phase transmission grating. It is therefore expected that phase transmission gratings and corrugated-surface reflection gratings have similar properties.

The diffraction efficiency of sinusoidal surface gratings as a function of λ / Λ depends on the modulation depth, or h/Λ ratio (h: peak to peak groove height), and on the angle of incidence [4.7, 8]. Small modulation depths ($h/\Lambda < 0.05$) can be treated by a scalar theory giving a peak diffraction efficiency of 34%, i.e., the same as a phase transmission grating. High modulation depths ($h/\Lambda > 0.25$) fall outside of the scalar diffraction regime. In these cases, the diffraction efficiency in one plane of polarization can theoretically reach 100 % while the efficiency for the orthogonal polarization is very low. Such high efficiencies are only realized for ratios $\lambda/\Lambda > 0.7$. Similar results are expected for transmission gratings with strong index modulation.

4.2.4 Thin Grating Condition

A grating with an arbitrary thickness d may be divided into thin elements (Fig. 4.4). The beam diffracted by the first grating element is advanced in phase by (λ: vacuum wavelength; n: refractive index)

$$Q/2 = (2\pi n/\lambda)\,(d - d\cos\phi_1) \tag{4.16}$$

relative to a beam diffracted by the exit grating element. Introducing the diffraction angle $\sin\phi_1 = \lambda/\Lambda n$, which is measured inside the material and assumed to be small ($\phi_1 \approx \sin\phi_1$), one obtains

$$Q = 2\pi d\lambda/\Lambda^2 n \ . \tag{4.17}$$

If the phase difference is sufficiently small ($Q \ll 1$) the beams from all the grating elements interfere constructively. If the phase difference is large ($Q \gg 1$), destructive interference occurs and the total diffracted intensity becomes small. Only in the case of oblique incidence at angle α and a diffraction angle $\phi_1 = 2\alpha$, which can be considered as reflection of the incident beams at the grating planes, there is no phase difference between all the beams diffracted at the grating planes (Fig. 4.4b). This case leads to the Bragg condition (2.46). Because all the waves diffracted at the thin grating elements interfere constructively, strong diffraction may be observed in the Bragg direction also with a large grating thickness.

a)   b)

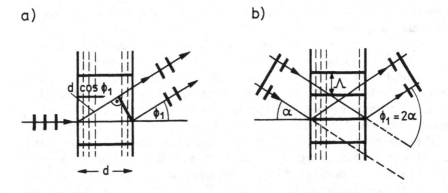

Fig. 4.4a, b. Diffraction at a thick grating treated by successive diffraction of the incident beam at thin grating elements. (a) Normal incidence; (b) Bragg diffraction

4.3 Coupled-Wave Theory for Thick Transmission Gratings

The general planar diffraction problem is depicted in Fig. 4.5 following [4.2, 9]. An electromagnetic wave is obliquely incident upon a slanted-fringe planar grating bounded by homogeneous media. In general, there will be simultaneously both forward-diffracted and backward-diffracted waves as shown in the figure.

The incident plane wave polarization is perpendicular to the plane of incidence (*H*-mode). The extension to *E*-polarization is treated in [4.9]. The relative permittivity (dielectric constant) in the grating region is given by

$$\varepsilon(x, z) = \varepsilon + \varepsilon_1 \cos\left(\boldsymbol{q} \cdot \boldsymbol{r}\right)$$

$$= \varepsilon + \varepsilon_1 \cos\left[q(x \sin \phi + z \cos \phi)\right] , \tag{4.18}$$

where ε_1 is the complex amplitude of the sinusoidal relative permittivity and $q = 2\pi/\Lambda$. The grating slant angle ϕ is introduced because a slant may appear in experiments if the grating material is not oriented perpendicularly to the interference fringes used for grating production. In addition, a slant angle of $\phi \approx \pi/2$ leads to reflection gratings which are thus covered in a unified description with transmission gratings.

The average relative permittivity ε may be complex and is related to the complex refractive index $\tilde{n}$ and real refractive index n and absorption constant K by

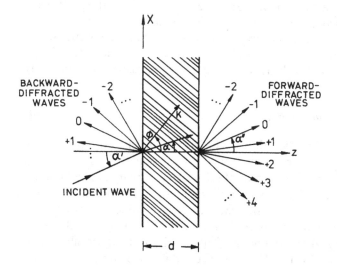

Fig. 4.5. Thick planar grating geometry. Reflection and refraction at the boundaries is neglected in the calculations ($\alpha' = \alpha = \alpha''$)

$$\tilde{n} = n + iK/2k \ , \tag{4.19}$$

$$\varepsilon = (\tilde{n})^2 = n^2 + inK/k - K^2/4k \ . \tag{4.20}$$

Here $k = 2\pi/\lambda$ and λ is the vacuum wavelength. The incident wave is given by

$$A_i = \frac{A_i}{2} \exp(i\mathbf{k}_i \cdot \mathbf{r}) + \text{c.c.} \tag{4.21}$$

$$= \frac{A_i}{2} \exp[ik_i(x \sin \alpha + z \cos \alpha)]$$

where α is the angle of incidence measured inside the grating material. Reflection of the incident and diffracted waves at the boundaries will not be treated here although it is possible to include reflection in a more complete discussion [4.2].

The generation and propagation of diffracted waves in the grating is described by the wave equation which is scalar for H-polarization and reduces to the Helmholtz equation if the time dependence is separated out:

$$\nabla^2 E + k^2 \varepsilon(x, z) E = 0 \ . \tag{4.22}$$

The field inside the grating is expressed as a superposition of plane waves with amplitudes S_m varying in z-direction

$$E(x, z) = \sum_{m=-\infty}^{\infty} S_m(z) \exp(i\boldsymbol{\sigma}_m \cdot \mathbf{r}) + \text{c.c.} \ . \tag{4.23}$$

The partial waves inside the modulation material are visualized as diffracted waves that progress through the planar slab and couple energy back and forth from and to each other as they progress. The sum $E(x, z)$ of the partial waves satisfies the wave equation. However, the partial waves do not individually satisfy the wave equation. The diffracted wavevectors $\boldsymbol{\sigma}_m$ inside the grating may be represented by

$$\boldsymbol{\sigma}_m = \mathbf{k}_i - m\mathbf{q} \quad \text{with} \quad m = 0, \pm 1, \pm 2 \ldots \ . \tag{4.24}$$

This representation is somewhat arbitrary [4.2]. It is sufficient to use only the x-component of (4.4) as can be expected from the discussion of diffraction at thin gratings in Sect. 4.2. The phase change of the diffracted wave in the z-direction is then exclusively described by the wave amplitude. It should be pointed out here that generally $\boldsymbol{\sigma}_m$ is the wave-vector of a diffracted wave inside the grating. The wave vectors outside the grating have to be formed by matching the field inside and outside the grating at the boundary. This has been discussed already in Sect. 4.2, leading to (4.6) for the direction of the diffracted waves.

Substituting (4.18, 23) into the wave equation (4.22) and performing the differentiations gives

$$\sum_m \left[\frac{\partial^2 S_m}{\partial z^2} + 2\,\mathrm{i}\,(k_\mathrm{i}\cos\alpha - mq\cos\phi)\,\frac{\partial S_m}{\partial z} - \sigma_m \cdot \sigma_m S_m \right.$$

$$\left. + k^2\varepsilon S_m + \frac{k^2\varepsilon_1}{2}\,S_{m-1} + \frac{k^2\varepsilon_1}{2}\,S_{m+1} \right] \exp\,(\mathrm{i}\sigma_m \cdot r) = 0 \ . \qquad (4.25)$$

This equation must be satisfied for all values of the variables. Thus the coefficient of each exponential must individually be zero. Using this and the definitions of $k_\mathrm{i} = nk$ and $k = 2\pi/\lambda$, the rigorous coupled-wave equations are obtained:

$$\frac{d^2 S_m}{dz^2} + 4\pi\mathrm{i} \left(\frac{n\cos\alpha}{\lambda} - \frac{m\cos\phi}{\Lambda} \right) \frac{dS_m}{dz} + \frac{4\pi^2 m}{\Lambda^2} \left[\frac{2n\Lambda}{\lambda}\cos\,(\alpha - \phi) - m \right] S_m$$

$$+ \frac{4\pi^2}{\lambda^2}\,(\varepsilon - n^2)\,S_m + \frac{2\pi^2\varepsilon_1}{\lambda^2}\,(S_{m+1} + S_{m-1}) = 0 \ . \qquad (4.26)$$

This is an infinite set of second-order coupled differential equations. The rigorous solution of these equation was discussed in [4.2]. Enormous simplifications can be achieved using suitable approximations. A number of famous analytic expressions occur for special limiting cases. Second derivatives may be neglected if it is assumed that the partial wave amplitudes $S_m(z)$ vary only slowly over distances compared to the wavelength [4.9]. In this case, the absorption must also be small and one obtains from (4.20) the approximation $\varepsilon - n^2 \approx \mathrm{i}nK/k$.

Exact Bragg incidence leads to a further interesting simplification of (4.26). For a slanted grating, the Bragg condition is derived as:

$$\cos\,(\alpha - \phi) = m\,\frac{\lambda}{2\sqrt{\varepsilon}\,\Lambda} \ , \qquad m = 0, \pm 1, \pm 2 \dots . \qquad (4.27)$$

For an unslanted transmission grating with $\phi = \pi/2$ and $\lambda_2 = \lambda/n$ this equation is equivalent to (2.46). Introducing (4.27) into (4.26), the first term with S_m disappears, but only for the mth partial wave for which the Bragg condition (4.27) is met. Similarly, the first S_0-term disappears in the equation for the transmitted wave ($m = 0$). The first S_m-term strongly influences the solution of (4.26). This is seen easily if the second derivative and absorption are neglected (i.e., $\varepsilon = n^2$) and the coupling term is set constant. Then (i) if the S_m-term is zero, the amplitude S_m builds up linearly $S_m \sim z$, (ii) if the S_m-term is present, S exhibits an oscillatory behaviour. The S_m-term in (4.26) is therefore called a dephasing term which prevents a continous build-up of S_m. Only for the primary light wave and a possible Bragg wave, (4.26) do not contain a dephasing term, so that the other waves may be neglected so as to result in the two-wave approximation.

Two-wave first-order coupled-wave theory has been first applied to holography by *Kogelnik* [4.9]. His 1969 paper is now widely referenced because of the comprehensive coverage of (1) absorption, phase and mixed gratings, (2) deviation from Bragg incidence, (3) transmission ($\phi = \pi/2$) and reflection ($\phi = 0$) gratings, (4) slanted fringe gratings and (5) both *H*- and *E*-polarisation. For an extensive discussion of the results, the reader is referred to the original paper [4.9]. Here only some simple cases shall be treated. If the Bragg condition is met for the first-order diffracted wave ($n = 0$), only S_0 and S_1 have to be considered and, for an unslanted grating ($\phi = \pi/2$), (4.26) reads approximately

$$\frac{dS_0}{dz} = -\frac{K}{2\cos\alpha} S_0 + i\frac{\pi\varepsilon_1}{2n\,\lambda\cos\alpha} S_1 \; ,$$

$$\frac{dS_1}{dz} = -\frac{K}{2\cos\alpha} S_1 + i\frac{\pi\varepsilon_1}{2n\,\lambda\cos\alpha} S_0 \; . \tag{4.28}$$

The solution of these equations with the initial conditions $S_0(z=0) = A_i$ and $S_1(z=0) = 0$ is

$$S_0 = A_i \exp\left(\frac{-Kz}{z\cos\alpha}\right) \cos\left(\frac{\pi\varepsilon_1 z}{2n\lambda\cos\alpha}\right)$$

$$S_1 = iA_i \exp\left(\frac{-Kz}{2\cos\alpha}\right) \sin\left(\frac{\pi\varepsilon_1 z}{2n\lambda\cos\alpha}\right) . \tag{4.29}$$

In the case of a pure phase grating, ε_1 is real, and (4.29) is directly applicable. Allowing complex values for ε_1, (4.29) is valid also for amplitude and mixed gratings.

Using the following relations between ε, ε_1 and the refractive index n and absorption coefficient K

$$\sqrt{\varepsilon} = \tilde{n} = n + iK\lambda/4\pi \; ,$$

$$\varepsilon_1 = \Delta\varepsilon \approx 2\tilde{n}\Delta\tilde{n} \approx 2n(\Delta n + i\lambda\Delta K/4\pi) \; , \tag{4.30}$$

one obtains for transmitted I_0 and diffracted I_1 intensities normalized to the incident intensity I_i:

$$\frac{I_0}{I_i} = \exp\left(\frac{-Kz}{\cos\alpha}\right)\left(\cos^2\frac{\pi\Delta n z}{\lambda\cos\alpha} + \cosh^2\frac{\Delta K z}{4\cos\alpha} - 1\right) , \tag{4.31}$$

$$\eta = \frac{I_1}{I_i} = \exp\left(\frac{-Kz}{\cos\alpha}\right)\left(\sin^2\frac{\pi\Delta n z}{\lambda\cos\alpha} + \sinh^2\frac{\Delta K z}{4\cos\alpha}\right) . \tag{4.32}$$

Equation (4.32) gives the diffraction efficiency η of a mixed grating in an absorbing material. A pure phase grating is obtained for $\Delta K = 0$, a pure

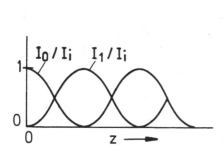

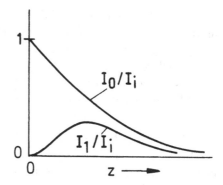

Fig. 4.6. Phase-grating: Normalized intensities of the primary wave I_0/I_i and diffracted wave I_1/I_i in dependence of the grating thickness z for the Bragg case

Fig. 4.7. Amplitude grating: Normalized intensities of the primary wave I_0/I_i and diffracted wave I_1/I_i in dependence of the grating thickness z for the Bragg case. Figure is not true to scale

absorption grating for $\Delta n = 0$. In Fig. 4.6, the normalized intensity of the two waves is shown for a pure phase grating with no absorption, i.e. $K = 0$. The intensity oscillates back and forth between the primary and diffracted beam. To achieve the maximum diffraction efficiency η of 100 %, the thickness of the grating has to be chosen to $d = \lambda \cos \alpha / \Delta n$. Equations (4.31 and 32) with $\Delta K = 0$ giving a periodic energy exchange between the two beams are called Pendellösung in x-ray and electron diffraction.

In Fig. 4.7, the normalized intensities for an absorption grating are shown. If negative absorption, i.e. optical gain, is excluded, the maximum value of ΔK is equal to K. The highest diffraction efficiency possible for an absorption grating is reached in the limiting case $\Delta K = K$ for a value of $Kz/\cos \alpha = 2 \ln 3$. According to (4.32), the maximum efficiency has a value of $(I_1/I_i)_{max} = 1/27$ or 3.7 %.

The zero- and first-order intensities of an absorption grating exhibit an interesting effect which has been long known in x-ray and electron diffraction as the Borrmann effect [4.10]. In its simplest version, this effect can be described by (4.31 and 32) with $\Delta n = 0$ and $\Delta K z / 4 \cos \alpha \gg 1$, which leads to

$$I_0 \approx I_1 \approx I_i \exp \left[-(K - \Delta K/2)z/\cos \alpha \right] . \tag{4.33}$$

The effective absorption constant $(K - \Delta K/2)$ appears to be reduced when compared to the average value K of the absorption constant. In the limiting case $K = \Delta K$, the effective absorption constant amounts only to $K/2$. The two beams in the Bragg case therefore may have intensities I_0, I_1 which are orders of magnitude larger then a beam which travels through the grating at an angle different from the Bragg angle and seeing the average absorption constant K. A thorough discussion of such Borrmann-like effects in volume holography has been given in [4.11].

There are other important features of thick gratings which can be obtained by more accurate solutions of the coupled-wave equations. For planning and interpreting experiments, deviations from the Bragg condition have to be considered. Such deviations may be caused by nonperfect angular incidence or by wavelength shifts or spread of the incident beam. The angular and wavelength sensitivity of various grating types was extensively discussed by *Kogelnik* [4.9].

4.4 Reflection Volume Gratings

Beams reflected from a grating can be produced by different physical mechanisms. First, reflected beams are produced by transmission gratings where the incident beam is reflected at the periodic boundary leading to first- and higher order reflected beams, as shown in Fig. 4.5. Similarly, the various diffracted beams propagating in a transmission grating are reflected at the second boundary. Diffraction at corrugated, reflecting surfaces used in spectroscopic apparatus is another example.

In this section, however, reflection gratings are considered which are similar to the transmission gratings discussed in Sect. 4.3 but have the grating planes $[\varepsilon(x, z) = \mathrm{const}]$ more or less parallel to the boundary. Unslanted reflection gratings are described by $\phi = 0$ in Fig. 4.5. The orientation of the grating planes and the Bragg condition (4.27) for this case are sketched in Fig. 4.8. The first-order wave is diffracted into the region in front of the grating. In the region behind the grating, only the undiffracted transmitted beam appears. This situation is similar to the reflection of a beam at the boundary between two materials with different refractive indices. Therefore a volume grating with $\phi \approx 0$ is called a reflection grating and the diffracted beams are often denoted as reflected beams.

The mathematical description of thick transmission gratings in Sect. 4.3 is valid also for reflection gratings. For exact Bragg incidence, the amplitudes of the transmitted and first-order reflected waves are derived from (4.26) with $\phi = 0$, neglecting second derivatives and assuming small absorption to:

$$\frac{dS_0}{dz} = -\frac{K}{2 \cos \alpha} S_0 + i \frac{\pi \varepsilon_1}{2 n \lambda \cos \alpha} S_1 , \tag{4.34}$$

$$\frac{dS_1}{dz} = +\frac{K}{2 \cos \alpha} S_1 + i \frac{\pi \varepsilon_1}{2 n \lambda \cos \alpha} S_0 . \tag{4.35}$$

These equations have to be solved with the boundary conditions $S_0(z=0) = A_i$ and $S_1(z=d) = 0$ leading to

$$S_1(z) = \frac{-i\varkappa A_i [\exp \gamma(z-d) - \exp -\gamma(z-d)]}{(\gamma - \delta) \exp(-\gamma d) + (\gamma + \delta) \exp(\gamma d)} \tag{4.36}$$

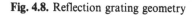

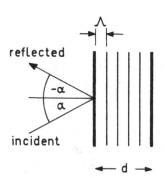

Fig. 4.8. Reflection grating geometry

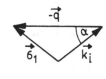

where

$$\varkappa = \pi \varepsilon_1 / 2 \, n_0 \lambda \cos \alpha$$

$$\delta = K/2 \cos \alpha \tag{4.37}$$

$$\gamma = \sqrt{\varkappa^2 + \varepsilon_2^2} \ .$$

The amplitude S_0 of the direct wave can be obtained by insertion of (4.36) into (4.35) and performing the differentiation.

The amplitude S_1 of the reflected wave at the boundary $z = 0$ of the grating is given by

$$S_1(0) = \frac{i \varkappa A_i}{\delta + \gamma \coth(\gamma d)} \ . \tag{4.38}$$

For a pure phase grating with $K = 0$, $\delta = 0$ and $\gamma = \varkappa = \pi \Delta n / \lambda \cos \alpha$, the diffraction efficiency is given by

$$\eta_p = \text{tgh}^2 (\pi \Delta n \, d / \lambda \cos \alpha) \ . \tag{4.39}$$

For sufficiently large values of the phase shift $\pi \Delta n \, d / \lambda \cos \alpha$, the diffraction efficiency approaches 100 %.

For a pure amplitude grating with $\varkappa = i \Delta K / 4 \cos \alpha$, $\delta = K/2 \cos \alpha$ and $\gamma = \sqrt{K^2 - (\Delta K/2)^2} / 2 \cos \alpha$ one obtains

$$\eta_a = \left(\frac{(\Delta K/2)}{K + \sqrt{K^2 - (\Delta K/2)^2} \coth [d \sqrt{K^2 - (\Delta K/2)^2} / 2 \cos \alpha]} \right)^2 \ . \tag{4.40}$$

For the deepest allowable modulation, where we have $K = \Delta K$, this equation predicts a maximum diffraction efficiency $\eta_{a,\text{max}} = (2 + \sqrt{3})^{-2} = 7.2 \%$ which is obtained for $d \Delta K \to \infty$. This means that either the thickness d or the amplitude ΔK of the absorption coefficient have to be large. For further discussion of the properties of reflection gratings, the reader is referred to [4.9].

4.5 Self-Diffraction

Self-diffraction denotes several interaction effects which are observed when two light beams of equal frequency intersect in a nonlinear material. The two beams form an interference field resulting in a periodic variation of the permittivity ε. This variation acts like a grating which diffracts the incident beams. An extensive survey of the resulting self-diffraction effects has been given in [4.12]. In the case of a thin grating, new beams appear, as shown in Fig. 4.9a. In the case of a thick grating, most of the diffraction orders are interference quenched. Self-diffraction may then be observed in the geometries shown in Fig. 4.9b–d. In Fig. 4.9b, c, Bragg self-diffraction of beam R into the direction of beam S and vice-versa is considered. The diffracted light of beam R (or S) interferes with beam S (or R) and changes the amplitude and phase of this beam. In materials with local and instantaneous response, i.e. the permittivity change ε_1 is strictly proportional to the light intensity in the interference region, only the phases of the two beams change but there is no energy exchange. Amplitude changes appear in materials with nonlocal or noninstantaneous response.

Nonlocal response means that the $\varepsilon(x, z)$-grating is spatially shifted with respect to the intensity grating. Such a shift appears in electrooptic crystals like $LiNbO_3$, $Bi_{12}SiO_{20}$ or $KNbO_3$ where light absorption produces a spatial distribution of electrons (or holes) in the conduction band. These electrons diffuse into the dark regions and are captured by deep traps. The resulting space-charge field modulates the refractive index via the electrooptic effect and the resulting grating can be shifted by a quarter of the period $\Lambda/4$ relative to the interference field, see Sect. 3.6. Another possibility is to produce an artificial mismatch between the grating and the interference field. This can be done by movement of the nonlinear material, by phase shifting one of the recording beams with respect to the other one or by application of electric and magnetic fields in the case of charge-carrier gratings [4.12].

Nonlocal response leads to an energy exchange between the two beams. For sufficiently large thickness the total energy of the two beams can be transferred into one beams as indicated in Fig. 4.9b.

Beam amplification or energy transfer is also possible if the ε_1-response is local but has a time lag compared to the intensity of intersection beams. Such a nonstationary response appears, e. g., during the build-up of a thermal grating by a laser pulse with a pulse-width short compared to the thermal relaxation time [4.12]. Since there is no prefered direction in a material with local response, energy transfer appears only when the intensity of the two intersecting beams is different and energy transfer always occurs from the strong to the weak beam.

Another possibility to observe self-diffraction with thick gratings is nonlinear phase-matching, as sketched in Fig. 4d. A weak beam and a strong one are incident on the thick nonlinear material where the grating is produced. The first non-Bragg diffraction order of the strong beam can be observed if the beam intensity of the strong beam is chosen correctly. This is understood by

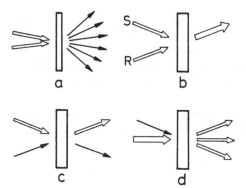

Fig. 4.9a–d. Self diffraction geometries: (a) Self-diffraction at a thin grating, (b) Bragg self-diffration of two beams in a medium with nonlocal responses, (c) non-stationary Bragg self-diffraction in a medium with local response leading to amplification of the weak beam, and (d) self-diffraction with nonlinear phase-matching of a non-Bragg diffraction order

considering the phase change in the diffracted beam produced by the nonlinear diffraction process. This phase change can compensate the geometrical dephasing of the non-Bragg diffracted beam. The intensity of this beam can therefore build up appreciably so that three beams appear behind the material. Experimental investigations of such three-beam self-diffraction effects are scarce so that this subject [4.12] is not outlined here in detail.

In [4.12] four-wave interactions are considered as additional self-diffraction effects. Because of the large importance and interest this subject has received in the literature, we discuss four-wave mixing effects separately in Sect. 4.6. Also, in four-wave mixing experiments, the grating build-up and detection can be attributed to different beams. Therefore in our terminology four-wave mixing if not a self-diffraction effect, where the grating is build-up and detected by the same beams. This indicates that the terminology used to describe self-diffraction is not yet finally established [4.12].

In the following, the self-diffraction effects sketched in Fig. 4.9a–c are described in more detail.

4.5.1 Self-Diffraction at Thin Gratings

Self-diffraction at thin thermal phase gratings has been discussed in [4.13] starting from Maxwell's wave equation coupled with the proper material equations. A set of nonlinear coupled integral-differential equations for the amplitudes of the diffracted waves was obtained. With suitable approximations, the diffracted intensity $I_{\pm m}$ into the mth order has been obtained, i.e.,

$$I_{\pm m} = TI_0 [J_m^2(\phi) + J_{m+1}^2(\phi)] \ , \tag{4.41}$$

where J_m and J_{m+1} are Bessel functions, I_0 is the incident intensity, and T the sample transmission. $\phi = 2\pi\Delta n d/\lambda$ is the time dependent phase amplitude of the refractive index grating. For small ϕ, only the Bessel function J_m has to be considered in (4.41), as was assumed in [4.14]. Equation (4.41) is easily interpreted by considering diffraction at a permanent phase grating where a single incident beam leads to an mth order diffracted intensity proportional to $J_m^2(\phi)$, see (4.14). In a self-diffraction arrangement (Fig. 4.9a), two incident beams are present. The mth diffraction order from one beam coincides with the $(m+1)$th diffraction order of the other beam. The diffracted intensity in a self-diffraction experiment is therefore determined by the sum of the diffracted intensities of the single beams.

Although (4.41) has been derived [4.13] only for thermal phase gratings, it seems obvious that it is also a useful approximation for self-diffraction at other types of thin amplitude and phase gratings.

4.5.2 Self-Diffraction at Thick Gratings with Constant Amplitude

Two coherent beams

$$R = R(z) \exp\left[ik(z \cos \alpha + x \sin \alpha)\right] ,$$
$$S = S(z) \exp\left[ik(\cos \alpha - x \sin \alpha)\right] \tag{4.42}$$

are considered symmetrically incident on a nonlinear material at an angle α relative to the z-axis. The two beams produce periodic variations of the permittivity

$$\varepsilon(x) = \varepsilon + \varepsilon_1 \cos(qx - \varphi) , \tag{4.43}$$

where $q = 2\pi/\Lambda = 2k \sin \alpha$. The phase angle φ is introduced to describe a possible shift of the permittivity grating with respect to the interference pattern of the two beams. From the coupled-wave analysis outlined in Sect. 4.3, we find for the amplitudes in analogy to (4.28) neglecting absorption ($K = 0$)

$$dR(z)/dz = i\varkappa \exp(-i\varphi) S(z) ,$$
$$dS(z)/dz = i\varkappa \exp(i\varphi) R(z) . \tag{4.44}$$

In the following the coupling constant $\varkappa$ is assumed to be real and is given by

$$\varkappa = \pi \varepsilon_1 / 2 n_0 \lambda \cos \alpha = \pi \Delta n / \lambda \cos \alpha , \tag{4.45}$$

where Δn is the amplitude of the corresponding refractive index grating and λ the vacuum wavelength.

For a first evaluation of (4.44), we assume that ε_1, $\varkappa$ and φ have reached steady state values which are independent of R and S. With the boundary conditions $R(0) = 1$, $S(0) = A$, the beam amplitudes are obtained as [4.15]

$$R(z) = \cos \varkappa z + iA \exp(-i\varphi) \sin \varkappa z , \tag{4.46}$$

$$S(z) = A \cos \varkappa z + i \exp(i\varphi) \sin \varkappa z , \tag{4.47}$$

and the intensities are given by

$$I_R = \cos^2 \varkappa z + A^2 \sin^2 \varkappa z - A \sin 2\varkappa z \sin \varphi ,$$

$$I_S = A^2 \cos^2 \varkappa z + \sin^2 \varkappa z + A \sin 2\varkappa z \sin \varphi .$$
(4.48)

We now consider first an unshifted grating with $\varphi = 0$. In this case, the intensities of the two beams change as one would expect by considering diffraction of the individual beams at a thick phase grating, see (4.31, 32) and adding the intensities of the transmitted and diffracted beams for each propagation direction. For equal incident intensities, i.e. $A = 1$, one obtains no change of the beam intensities in the case $\varphi = 0$.

For equal incident beam intensities $(A = 1)$, an intensity transfer between the beams is only obtained for $\varphi \neq 0$:

$$I_R = 1 - \sin \varkappa z \sin \varphi ,$$

$$I_S = 1 + \sin \varkappa z \sin \varphi .$$
(4.49)

The intensity is transferred to the incident beams towards which the grating is shifted. Maximum intensity or energy transfer occurs for $\varphi = \pi/2$, i.e. the grating has to be shifted by $\Lambda/4$.

The φ-dependence (4.48 or 49) of the transmitted beams can be explained as follows. When self-diffraction takes place, two collinear waves propagate in the direction of each of the interacting beams, namely the transmitted zero-order wave from one of the beams and the first-order diffraction wave from the other. For the unshifted grating, these two waves are phase shifted by $\pi/2$ relative to each other, so that the intensities are added. When the grating is shifted by a quarter of period, there is an additional phase difference of $\pm \pi/2$ between the waves. For the acceptor beam direction, the two waves are then in phase and add constructively, whereas for the donor beam they are in antiphase and add destructively. This gives the possibility to combine the intensities of the two interacting beams into a single beam. The other beam may be quenched completely by interference.

Another important conclusion which can be drawn from (4.46, 7) is that, for $\varphi = \pi/2$, the beam amplitudes $R(z)$ and $S(z)$ are real which means that the interference fringes are parallel to the z-axis. On the other side, for $\varphi = 0$, the amplitudes $R(z)$ and $S(z)$ contain an additional complex phase which leads to bending of the interference pattern. In this case, the interference fringes are no more parallel to the z-axis. It follows that the grating also bends in the nonlinear material and (4.43) is not valid anymore. A more careful analysis is therefore needed taking into account the grating build-up by the interference pattern.

4.5.3 Self-Diffraction at Thick Gratings with Intensity Dependent Amplitude

We assume now that the refractive index change ε_1 is proportional to the light intensity in the interference pattern of R and S, so that

$$\varepsilon_1 \propto a R(z) S^*(z) \exp i(qx - \varphi) + \text{c.c.} , \qquad (4.50)$$

where a is a coupling parameter. The phase angle φ is again introduced to describe a possible shift of the ε_1-grating with respect to the interference pattern. From the coupled wave analysis, one obtains now in analogy to (4.44):

$$dR(z)/dz = ia \exp(-i\varphi) SS^*R ,$$
$$\qquad\qquad\qquad\qquad\qquad\qquad\qquad (4.51)$$
$$dS(z)/dz = ia \exp(i\varphi) RR^*S .$$

These equations will now be solved by assuming an unshifted grating ($\varphi = 0$) and a shifted grating with $\varphi = \pi/2$.

Unshifted Grating ($\varphi = 0$)
With $R(z) = |R| \exp(i\varphi_R)$ and $S(z) = |S| \exp(i\varphi_S)$, one obtains from (4.51)

$$\frac{d|R|}{dz} = 0 , \quad \frac{d|S|}{dz} = 0 , \qquad (4.52)$$

$$\frac{d\varphi_R}{dz} = aSS^* , \quad \frac{d\varphi_S}{dz} = aRR^* . \qquad (4.53)$$

Equations (4.52) imply that the amplitudes $|R|$ and $|S|$ of the two beams do not change in the nonlinear grating material. There is no energy transfer between the two beams for arbitrary incident intensities. From (4.53), the phase difference between the two beams is obtained as

$$\varphi_S - \varphi_R = a(|R|^2 - |S|^2)z . \qquad (4.54)$$

This phase difference leads to a slant of the permittivity grating which is now given by

$$\varepsilon_1 = 2a \cos[qx - (\varphi_S - \varphi_R)] . \qquad (4.55)$$

The slant angle ϕ is given by

$$\text{tg}\,\phi = a(|R|^2 - |S|^2)/q . \qquad (4.56)$$

We conclude that self-diffraction at a thick grating with an amplitude proportional to the light intensity does not result in an energy transfer between the two beams. Only the phases of the beams change, resulting in a grating slant if the incident intensities $|R|^2$ and $|S|^2$ of the beams are different.

Shifted Grating ($\varphi = \pi/2$)
Equations (4.51) read now

$$\frac{dR}{dz} = aSS^*R \ , \tag{4.57}$$

$$\frac{dS}{dz} = -aRR^*S \ . \tag{4.58}$$

The phases of R and S remain unchanged. The permittivity grating is therefore not slanted. From (4.57, 58), one obtains for the intensities I_R and I_S of the two beams

$$\frac{dI_R}{dz} = 2\, aI_S I_R \ , \tag{4.59}$$

$$\frac{dI_S}{dz} = 2\, aI_S I_R \ . \tag{4.60}$$

The solution of these equations is given by

$$I_R = I_0/\{1 + [I_S(0)/I_R(0)]\exp(2\, aI_0 z)\} \ , \tag{4.61}$$

$$I_S = I_0/\{1 + [I_R(0)/I_S(0)]\exp(-2\, aI_0 z)\} \ , \tag{4.62}$$

where $I_0 = I_R(0) + I_S(0) = I_R(z) + I_S(z)$ is the total intensity of the incident beams. Equation (4.62) shows that, for sufficiently large thickness z of the nonlinear material, the total incident intensity can be transferred to the beam I_S.

4.5.4 Energy Transfer by Nonstationary Gratings

Energy transfer between two beams by self-diffraction in a nonlinear medium with local response does not take place with stationary (cw) intensity of the incident beams, see (4.52). Energy transfer is observed, however, in the case of pulsed excitation in materials in which the build-up time of the permittivity changes is comparable or larger than the pulse-width.

A qualitative understanding of the effect can be obtained with the help of Fig. 4.10. At the beginning of the pulse, the interference fringes are parallel to the bisectrix of the two beams and normal to the surface of the nonlinear material. The permittivity grating also builds up perpendicularly to the surface (Fig. 4.10a). During the following part of the pulse, diffraction at the grating leads to a phase shift of the two beams, compare (4.53). Consequently, the interference fringes become slanted. Because of the finite build-up time of the grating, the slant angle of the interference fringes is smaller than that of the grating planes (Fig. 4.10b). The corresponding shift between the interference

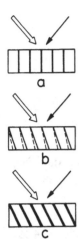

Fig. 4.10a–c. Development of the slant of the interference pattern (*solid lines*) and of the grating (*broken lines*) during nonstationary self-diffraction. (**a**) Beginning of the pulse, $t = 0$. (**b**) Interaction time comparable to grating buildup time, $t \approx \tau$. (**c**) Stationary state, $t \gg \tau$. After *Vinetskii, Kukhtarev, Odulov, Soskin* [4.12]

pattern and the gratings leads to an energy transfer from the strong beam to the weak beam as described by (4.51) with $\varphi \neq 0$. The weak beam amplication ceases in the steady state where the interference pattern and the grating are in phase (Fig. 4.10c).

A mathematical description of nonstationary self-diffraction has been given in [4.16]. An analytic solution in closed form is complicated, so that approximations for small gain of the weak beam or computer calculations are used [4.12].

4.6 Four-Wave Mixing

Four-wave mixing generally denotes the interaction of four light waves with different frequencies $\omega_1 \ldots \omega_4$ and propagation directions $k_1 \ldots k_4$. The interaction is due to a third-order nonlinear polarization of the material leading to a number of different effects, like third-harmonic generation and Raman-type frequency mixing. These effects have been investigated extensively [4.17] since the pioneering work of *Bloembergen* and co-workers [4.18], and *Maker* and *Terhune* [4.19].

In degenerate four-wave mixing, the frequencies of the incident light waves are equal ($\omega_1 = \omega_2 = \omega_3 = \omega_4 = \omega$) and the wave vectors are mutually antiparallel $k_1 = -k_2$, $k_3 = -k_4$. This is only one possible form of degeneracy, so that the term degenerate four-wave mixing is not fully specific. Therefore in this section, as in many other papers, the specification degenerate is omitted.

In the simplest but experimentally most important case of four-wave mixing, only two pump waves with $\omega_1 = \omega_2 = \omega$ and $k_1 = -k_2$, and a signal wave with $\omega_4 = \omega$ and k_4 arbitrary are incident on the nonlinear material (Fig. 4.11). Due to the third-order polarization, a reflected wave with $\omega_3 = \omega$ is created which has the opposite direction $k_3 = -k_4$ to the signal wave. If an additional fourth wave with $k_3 = -k_4$ is incident, this wave can be amplified.

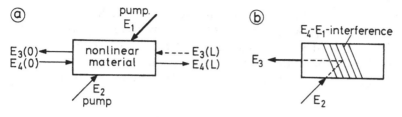

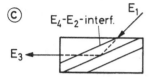

Fig. 4.11a–c. Grating interpretation of four-wave mixing

The four-wave mixing process can be understood also with the grating picture. The incident waves E_4 and E_1 interfere (Fig. 4.11b) and induce a grating with a vector $q = k_4 - k_1$. Bragg diffraction of E_2 leads to the wave E_3 with $k_3 = k_2 - q = -k_4$. A second contribution to the wave E_3 comes from the grating that is produced by interference of E_4 and E_3 and diffracts the wave E_1 (Fig. 4.11c). In the case of instantaneous, local and linear response of the material, the two contributions to the reflected wave are equal and it is sufficient to consider only the grating which is produced by interference of E_4 and E_1.

4.6.1 Four-Wave Mixing Described by Grating Interaction

For a theoretical analysis, the four waves ($m = 1, 2, 3, 4$) are defined by

$$E_m(r, t) = \tfrac{1}{2} A_m(z) \exp\left[i(k_m \cdot r - \omega t)\right] + \text{c.c.} \quad . \tag{4.63}$$

The relative permittivity modulation produced by interference of E_4 and E_1 is given by (SI units)

$$\varepsilon(x, z) = \varepsilon + \chi^{(3)} A_1 A_4^* \exp\left[i(k_1 - k_4) \cdot r\right] + \text{c.c.} \quad . \tag{4.64}$$

Here ε is the average relative permittivity connected to the refractive index n and absorption coefficient K. The development of the amplitudes $A(z)$ is given by the wave equation

$$\nabla^2 E + \varepsilon(x, z) k^2 E = 0 \quad , \tag{4.65}$$

with $k = 2\pi/\lambda$ (λ: vacuum wavelength).

Introducing (4.63, 64) into (4.65) yields

$$\frac{dA_3}{dz} = -\frac{ik}{2n}(\varepsilon - n^2) A_3 - \frac{ik}{2n} \chi^{(3)} A_1 A_2 A_4^* \quad . \tag{4.66}$$

In deriving (4.66), only terms with the factor $\exp[i(\mathbf{k}_3 \cdot \mathbf{r} - \omega t)]$ are considered and the relation $\mathbf{k}_3 = \mathbf{k}_1 + \mathbf{k}_2 - \mathbf{k}_4 = -\mathbf{k}_4$ is used. A_3 is assumed to vary slowly with z so that the second derivative may be neglected. The z-direction is parallel to $\mathbf{k}_4$. The absolute values of the wave-vectors are given by $k_m = nk$, where n is the refractive index. Using $\varepsilon - n^2 \approx inK/k$, where K is the absorption constant, one obtains

$$\frac{dA_3}{dz} = \frac{K}{2} A_3 - i\kappa A_4^* . \tag{4.67}$$

Here the coupling constant $\kappa = (\pi/\lambda n)\chi^{(3)} A_1 A_2$ is introduced. Note that (4.67) is often given with a positive sign in the literature. This is due to a different definition of the field amplitude, using the complex conjugate of our field amplitude given by (4.63).

When a wave E_3 is present in the nonlinear material, this wave produces additional gratings by interference with E_1 (and E_2) leading to diffracted waves in the direction of $\mathbf{k}_4$. The amplitude change dA_4/dz is derived similarly as dA_3/dz leading to

$$\frac{dA_4}{dz} = -\frac{K}{2} A_4 + i\kappa A_3^* . \tag{4.68}$$

4.6.2 Four-Wave Mixing Described by Nonlinear Polarization

The waves E_1, E_2, E_4 given by (4.63) induce a nonlinear polarization of the form [4.20]

$$P^{NL}(\omega = \omega + \omega - \omega) = \frac{\varepsilon_0}{2} \chi^{(3)} A_1 A_2 A_4^* \exp[i(\omega t + k_3 z)] + c.c. . \tag{4.69}$$

Because $\mathbf{k}_1 + \mathbf{k}_2 - \mathbf{k}_4 = \mathbf{k}_3$, the polarization wave P^{NL} has a constant phase related to the wave E_3, i.e. P^{NL} and E_3 are phase matched. The amplitude change of the electric field E_3 is given by the inhomogeneous wave equation (Si units)

$$\nabla^2 E_3 - \frac{\varepsilon}{c^2} \frac{\partial^2 E_3}{\partial t^2} = \frac{1}{\varepsilon_0 c^2} \frac{\partial^2}{\partial t^2} P^{NL} . \tag{4.70}$$

The combination of (4.63, 69 and 70) gives (4.67) if the same approximations as above are used. Similarly (4.68) is obtained.

Starting from the nonlinear polarization therefore leads to the same basic equations for four-wave mixing as obtained from the grating picture (Sect. 4.6.1).

4.6.3 Amplitudes of Reflected and Transmitted Waves

The solution [4.20] of the coupled equations (4.67, 68) with the assumption of undepleted pump intensities, i.e. constant $\varkappa$ and neglecting absorption ($K=0$), is given by

$$A_3(z) = \frac{\cos|\varkappa|z}{\cos|\varkappa|L} A_3(L) - i \frac{\varkappa \sin|\varkappa|(z-L)}{|\varkappa| \cos|\varkappa|L} A_4^*(0) , \tag{4.71}$$

$$A_4(z) = i \frac{|\varkappa| \sin|\varkappa|z}{\varkappa^* \cos|\varkappa|L} A_3^*(L) + \frac{\cos|\varkappa|(z-L)}{\cos|\varkappa|L} A_4(0) . \tag{4.72}$$

Here $A_4(0)$ and $A_3(L)$ are the amplitudes of the signal and reflected wave at their respective input planes ($z=L$, $z=0$). In the following discussion, only a single input wave with $A_4(0)$ at $z=0$ and $A_3(L)=0$ is considered. The reflected wave is given by

$$A_3(0) = \left(\frac{i\varkappa}{|\varkappa|} \tan|\varkappa|L \right) A_4^*(0) . \tag{4.73}$$

The reflected field $A_3(0)$ is thus proportional to the complex conjugate of the incident field $A_4(0)$. This important property of four-wave mixing is called phase-conjugation, as will be discussed later.

Equation (4.73) shows that, for $\pi/4 < |\varkappa|L < 3\pi/4$, the reflected wave intensity exceeds that of the input wave intensity. The signal wave is always amplified according to

$$A_4(L) = \frac{A_4(0)}{\cos|\varkappa|L} . \tag{4.74}$$

The powers of the two beams are given by

$$P_3(z) = P_4(L) \sin^2|\varkappa|(z-L) , \tag{4.75}$$

$$P_4(z) = P_4(L) \cos^2|\varkappa|(z-L) . \tag{4.76}$$

The total power $P_3 + P_4$ is constant in the material

$$P_4(L) = P_4(0)/\cos^2|\varkappa|L = P_3(z) + P_4(z) . \tag{4.77}$$

The power distribution among the signal and reflected waves is shown in Fig. 4.12 for a signal beam amplification $P_4(L)/P_4(0) = 1/\cos^2|\varkappa|L = 6$. Similar curves are obtained for other values of $|\varkappa|L$ with $\pi/4 < |\varkappa|L < \pi/2$.

When $|\varkappa|L = \pi/2$, the amplification of the signal and reflected waves becomes infinite [$P_4(L) = P_3(0) \rightarrow \infty$]. In this case, the two waves can build up from an initial small fluctuation to a large amplitude which is limited, e.g. by pump

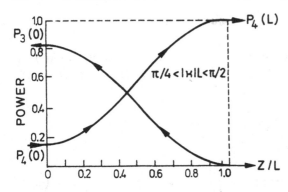

Fig. 4.12. Calculated distribution of signal P_4 and reflected power P_3 in a four-wave mixing device. The coupling parameter $|\kappa|L$ is chosen to give amplification of the reflected beam. After *Yariv* [4.20]

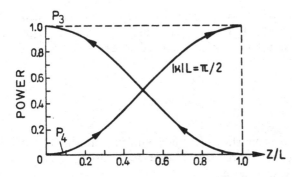

Fig. 4.13. Calculated power distribution in a four-wave mixing oscillator. After *Yariv* [4.20]

depletion. This corresponds to oscillation of the device without mirror feedback. The expected power distribution within such a four-wave mixing oscillator is shown in Fig. 4.13 where it is assumed that the total power $P(L)$ has reached its limiting value but (4.75) and (4.76) are still valid.

4.6.4 Further Development of the Theory

In the preceeding sections, the theory of degenerate four-wave mixing has been discussed only in its simplest version. In the literature, more detailed discussions are available including absorption [4.21, 22] in the nonlinear material, polarization effects [4.23–25], nonlinear index changes [4.26] and the dynamics [4.27–30] of the four-wave mixing interaction taking into account the time dependence of the fields and the nonlinear interaction.

4.6.5 Materials and Light Sources

Degenerate four-wave mixing, resulting in phase-conjugation, has been observed with a number of different materials in combination with suitable laser light sources. A general discussion of nonlinear materials, where grating and four-wave mixing effects can be produced, is given in Chap. 3. Special material aspects are discussed in the following references:

- Gases, vapours, plasmas (4.23, 31–33]
- Organic liquids and dye solutions [4.29, 34–36]
- Semiconductors (e. g., Si) [4.37]
- Photorefractive crystals [4.38–42]
- Four-wave mixing with CO_2-lasers [4.32, 43]
- Four-wave mixing with picosecond pulses [4.35, 36]
- Four-wave mixing of surface waves [4.44]

4.6.6 Applications

A great number of papers [4.20, 25, 35, 38–42, 45] on four-wave mixing deals with applications in information processing and distortion correction. Aspects of four-wave mixing for phase-conjugation are referenced in [4.45]. Only an arbitrarily selected number of references has been given here because this subject is outlined in more detail in Chap. 6.

Other possible applications discussed in Chap. 7 are narrow optical band-pass filters [4.46] and bistable optical devices [4.47].

In science (Chap. 5) degenerate four-wave mixing has great capability in high-resolution spectroscopy as was first experimentally recognized by *Liao* and colleagues in sodium [4.31]. Using a tunable narrow-band cw dye laser they measured sub-Doppler line shapes. Further spectroscopic aspects of four-wave-mixing have been discussed in [4.25, 48–51], see also Sect. 7.8.

4.7 Moving Gratings

Two laser beams with different frequencies $\omega_1 \neq \omega_2$, wave vectors k_1 and k_2 and electric field strengths

$$E_j = (A_j/2) \exp\left[i(k_j \cdot r - \omega_j t)\right] + \text{c.c.} , \quad j = 1, 2 \tag{4.78}$$

produce a nonstationary intensity distribution given by

$$I = \frac{1}{Z} \overline{(E_1 + E_2)^2} . \tag{4.79}$$

Here Z is the wave resistance of the material and the bar denotes time averaging. From (4.78) and (4.79), one obtains

$$I = \frac{1}{2Z} \left(|A_1|^2 + |A_2|^2 + A_1 A_2^* \exp i\left[(k_1 - k_2) \cdot r - (\omega_1 - \omega_2)t\right] + \text{c.c.}\right) . \tag{4.80}$$

In deriving this equation, averaging has been performed over times which are long compared to the optical periodes $T_{1,2} = 2\pi/\omega_{1,2}$ but short compared to the period $2\pi/(\omega_1 - \omega_2)$ corresponding to the difference frequency.

The intensity I exhibits a wavelike modulation with a grating vector

$$q = k_1 - k_2 \tag{4.81}$$

and a frequency

$$\Omega = \omega_1 - \omega_2 \; . \tag{4.82}$$

The frequencies ω_1, ω_2 should be defined so that $\Omega > 0$. Then the grating vector q is given without any ambiguity which exists for stationary gratings, compare (2.8). For $\omega_1 = \omega_2$ a stationary interference pattern is obtained as discussed in the previous sections of this volume.

The wave-like modulated light intensity (4.80) changes the optical properties of materials in the interference region resulting in a moving grating structure [4.52]:

$$\Delta\varepsilon = \Delta\chi = \chi^{(3)} A_1 A_2^* \exp i\left[(k_1 - k_2) \cdot r - (\omega_1 - \omega_2)t\right] + \text{c.c.} \; . \tag{4.83}$$

$\chi^{(3)}$ is the third-order nonlinear susceptibility.

Equation (4.83) is useful for steady-state conditions. If time-dependent effects are involved, the coupling between the light field and the permittivity may be much more complicated.

The gratings can be detected by diffraction of a third laser beam with a wave vector k_3 and a frequency ω_3. The directions and amplitudes of the diffracted waves can be obtained in analogy to Sect. 4.2 for the case of thin gratings, or Sect. 4.3 for the case of thick gratings. The directions of the diffracted waves are given in the following [4.53]. In addition it has to be considered that the frequencies of the diffracted beams are shifted by integer multiples of the grating frequency Ω. These frequency shifts are well known from light diffraction at ultrasonic waves and can be interpreted as a Doppler effect due to the motion of the grating, which acts as the source of the diffracted waves [4.54].

4.7.1 Thin Gratings

With a thin grating, diffraction is obtained for beams incident at arbitrary angles. The x-component $k_{4x}^{(m)}$ of the wave-vector of a mth-order diffracted beam is given in analogy to (4.5) by

$$k_{4x}^{(m)} = k_{3x} + mq \; , \quad m = 0, \pm 1, \pm 2, \pm 3 \ldots \; . \tag{4.84}$$

The x-direction is parallel to the grating vector q. The corresponding frequencies $\omega_4^{(m)}$ which determine also the absolute values $k_4^{(m)}$ of the diffracted wave vectors are given by

$$\omega_4^{(m)} = \omega_3 + m\Omega \; . \tag{4.85}$$

The diffraction angles can then be expressed by

$$\sin \phi_m = k_{4x}^{(m)}/k_4^{(m)} .$$ (4.86)

Due to the frequency shift of the diffracted beams, the diffraction angles also change when compared to a stationary grating.

4.7.2 Thick Gratings

The diffraction is treated by solving the time-dependent wave equation for the total electric field strength

$$\nabla^2 E - \frac{\varepsilon + \Delta\varepsilon}{c^2} \frac{\partial^2 E}{\partial t^2} = 0 .$$ (4.87)

The solution can be performed in a similar way, as for stationary gratings in Sect. 4.3. Optimum coupling between the incident and diffracted waves is obtained if the Bragg condition is obeyed, which for first-order diffraction is given by

$$k_4^\pm = k_3 \pm q ,$$ (4.88)

The absolute values of the wave-vectors are different ($k_4 \neq k_3$) because also the frequencies change

$$\omega_4^\pm = \omega_3 \pm \Omega .$$ (4.89)

The Bragg-condition is fulfilled for a set of incident wave-vectors k_3 forming two cones with q as axis. The diffracted wave-vectors form two corresponding cones.

Equations (4.85, 89) multiplied by Planck's constant $\hbar$, may be considered as the expression of energy conservation for the impact of photons (ω_3) with quasi-particles (Ω) representing the moving grating whereby photons (ω_4) are produced. Equations (4.84, 88) express momentum conservation. In the case of a thin grating, only the momentum in the direction of q is well defined. The transverse component of the momentum is given by a broad distribution in the momentum space. The width of this distribution is inversely proportional to the thickness of the grating. Therefore in the case of a thin grating, only the y-component of the wave vector is determined by momentum conservation. The other component may be derived from energy conservation.

4.7.3 Description by a Nonlinear Polarization

The interaction of the three incident waves E_1, E_2, E_3 which has just been discussed in the grating picture, may be alternatively described by a nonlinear polarization which acts as a source for the various diffracted waves. The total nonlinear polarization is composed of several terms, each radiating into the direction of a different diffraction order.

As an example, first-order Bragg-diffraction is considered with $\omega_4 = \omega_3 \pm \Omega$. . The nonlinear polarization term at this frequency is given by

$$P^{NL}(\omega_4 = \omega_1 - \omega_2 + \omega_3) = \frac{\varepsilon_0}{2} \chi^{(3)}(\omega_4 = \omega_1 - \omega_2 + \omega_3) A_1 A_2^* A_3$$

$$\cdot \exp i[(k_1 - k_2 + k_3) \cdot r - (\omega_1 - \omega_2 + \omega_3)t] \ . \quad (4.90)$$

This polarization radiates a light wave with a frequency ω_4. The coupling between the polarization wave and the light wave is most efficient if the phase-matching condition $k_4 = k_1 - k_2 + k_3$ is obeyed. This condition is equivalent to the Bragg condition (4.88) obtained in the grating picture. The amplitude of the Bragg-diffracted wave can be obtained from the inhomogeneous wave equation

$$\nabla^2 E - \frac{\varepsilon}{c^2} \frac{\partial^2 E}{\partial t^2} = \frac{1}{\varepsilon_0 c^2} \frac{\partial^2}{\partial t^2} P^{NL} \ . \quad (4.91)$$

Using the definitions (4.83, 90) of $\Delta\varepsilon$ and P^{NL}, the solutions of (4.91, 87) are equivalent. Of course, the same approximations have to be performed in solving the corresponding equations.

The equivalence between the grating picture and the nonlinear polarization description allows to interpret a number of nonlinear optical effects, which have been observed with noncollinear laser beams, as diffraction at moving gratings. Such effects are, e.g., (nondegenerate) four-wave mixing or Coherent Anti-Stokes Raman Scattering (CARS). The newly developed coherent scattering techniques (CARS, etc.) [4.55] may be considered as light diffraction at coherent molecular vibrations driven by a moving interference pattern. Detailed accounts of the theoretical and experimental aspects of four-wave mixing spectroscopy as well as related techniques have been summarized in several review articles [4.56, 57].

5. Investigation of Physical Phenomena by Forced Light Scattering

Forced light scattering denotes the diffraction of a light wave at a laser-induced grating. This term has been coined in analogy to classical spontaneous light scattering (Sect. 1.1.6). By measuring the diffraction efficiency, forced light scattering gives information on refractive-index changes and on the magnitude of the corresponding optically produced material excitations. Examples have been given already extensively in Chap. 3 and some others are added in the following, e.g., in Sect. 5.9. Measurements of this kind require calibrated intensities of the pump and probe beams.

Another kind of measurement performed in forced light scattering experiments is the observation of the time dependence of the diffraction efficiency. This gives information on the dynamics of the material excitation. The dynamical behaviour is easier to measure since a time-resolved record of the scattered intensity in arbitrary units is sufficient; alternatively a frequency spectrum will give the desired information.

The dynamics of the diffraction efficiency can be investigated either by pulsed excitation and direct-time-resolved observation, or by modulated excitation and measurement of the amplitude of the response as a function of the modulation frequency.

Frequency domain measurements can be done by two methods. First, the excitation beams are modulated in amplitude, which corresponds to standing-wave excitation. Second, excitation beams with a frequency offset are used, which results in a traveling wave excitation. In both cases, the excitation at a fixed point is temporarily modulated. The difference is that the phase of the excitation is constant for a standing wave and a linear function of position for traveling-wave excitation. Amplitude modulation of the excitation beams is useful for observation of comparatively slow effects, e.g. thermal gratings, and can be conveniently combined with phase-sensitive detection methods for weak diffracted signals. Traveling-wave excitation is promising for investigation of fast relaxation phenomena.

In the following Sect. 5.1, different methods for time resolved measurements are outlined. Applications of these techniques are described in Sects. 5.2–9.

5.1 Techniques for Investigating Grating Dynamics

The temporal evolution of laser-induced gratings can be studied in a straight-forward manner by inducing the gratings with a short laser pulse and by

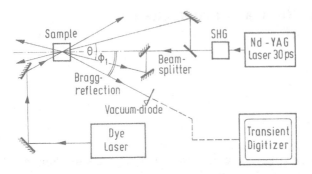

Fig. 5.1. Experimental setup for decay time measurement of a laser induced grating by diffraction of a quasi cw laser beam

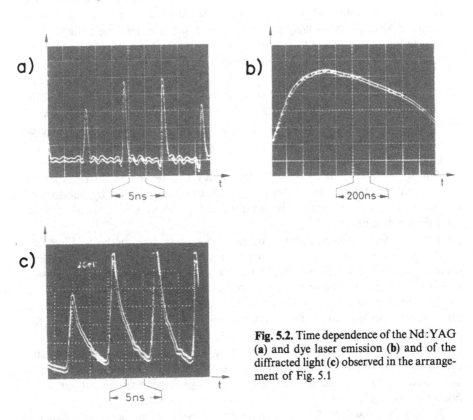

Fig. 5.2. Time dependence of the Nd:YAG (**a**) and dye laser emission (**b**) and of the diffracted light (**c**) observed in the arrangement of Fig. 5.1

observing the time dependence of a diffracted cw light beam. A typical arrangement [5.1] is shown in Fig. 5.1. The gratings are produced in this example by 30 ps pulses of a Nd:YAG laser which is frequency doubled by a SHG crystal. A beam splitter produces two waves which cross at an angle θ in the sample and build up the grating.

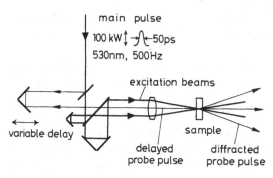

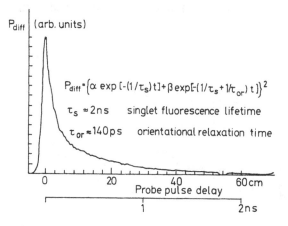

$$P_{\mathrm{diff}} \cdot \left\{ \alpha \exp\left[-(1/\tau_s)\,t\right] + \beta \exp\left[-(1/\tau_s + 1/\tau_{or})\,t\right] \right\}^2$$

$\tau_s \approx 2\,\mathrm{ns}$ singlet fluorescence lifetime

$\tau_{or} \approx 140\,\mathrm{ps}$ orientational relaxation time

Fig. 5.4. Time dependence of power P_{diff} diffracted at a grating induced in a 10^{-4} M solution of Rhodamine 6G in methanol [5.2a]

Physical mechanisms for the grating build up have been discussed already in Chap. 3. The grating is detected by diffraction of a quasi cw dye laser beam. Figure 5.2 shows the time dependence of the Nd:YAG laser emission, the dye laser emission, and the Bragg-diffracted light. The time dependence of the spatial amplitudes Δn, ΔK representing the grating is given approximately, according to (2.47), by the square root of the diffracted intensity. In general, the diffracted intensity is connected with Δn and ΔK in a more complicated manner, as is outlined in Chap. 4.

Time-resolved measurements with pulse excitation and diffraction of a quasi cw laser beam are possible down to a time resolution of approximately 1 ns if photodiodes and fast oscilloscopes are used for the detection of the diffracted light. It is possible to extend the time resolution down to several ps by using a streak camera. A less complex apparatus [5.2a] for this time region is sketched in Fig. 5.3. The main laser beam is used to produce two excitation beams which induce the grating in the sample. A delayed probe pulse derived from the main pulse is diffracted at the grating. The diffracted power P_{diff} versus the pulse probe delay is shown in Fig. 5.4. The decay process has been fitted by the sum of two exponentials with decay times $\tau_s = 2$ ns and $\tau_{or} = 140$ ps. These times have been identified as the singlet fluorescence lifetime and the orientational relaxation time of the dye solution Rhodamine 6G in methanol in which the grating has been produced. A detailed discussion of the grating mechanism is given in Sect. 3.4.

The sampling method (Fig. 5.3) is useful for the measurement of grating decay times down to 10 fs. This limit is given by the shortest laser pulses presently available. Even shorter decay times should be measurable by using moving gratings [5.2b]. A moving interference pattern is produced by superposition of two light waves with different wavelengths, as discussed in Sect. 4.7.

An experimental arrangement [5.3] suitable for the performance of experiments with moving gratings is shown in Fig. 5.5. This arrangement has been used originally for third-order light mixing experiments which can be understood, however, also with help of the grating picture. Two dye lasers are tuned to the

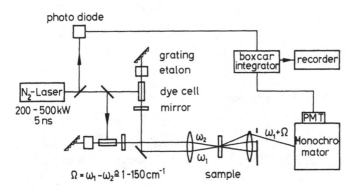

Fig. 5.5. Dual dye laser system for excitation of moving gratings and study of ultrafast relaxation processes [5.3]

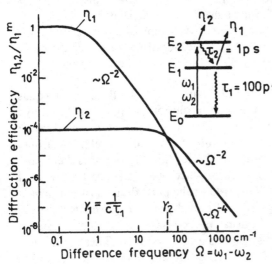

Fig. 5.6. Diffraction efficiencies η_1 and η_2 calculated for a three level model, plotted vs the frequency Ω of the moving grating [5.2b]. η_1 is obtained with a probing frequency $\omega_3 = (E_1 - E_0)/\hbar$ and η_2 with $\omega_3' = (E_2 - E_0)/\hbar \approx \omega_1 \approx \omega_2$

frequencies ω_1 and ω_2 with variable frequency difference Ω. The two beams with ω_1 and ω_2 induce a moving grating in the sample. The beam with ω_1 is self-diffracted with a frequency shift $+\Omega$. The power of the diffracted beam is measured as a function of the frequency Ω. If the frequency Ω is large compared to the inverse lifetime of the material excitation representing the grating, the material is not able to follow the rapid oscillations of the light energy in the interference region. The grating amplitude and the diffraction efficiency then decrease.

Experimental results using the moving grating technique have been scarce up to now. Therefore, as an example, the dependence of the diffraction efficiency on the difference frequency Ω is given in Fig. 5.6 as calculated in the original proposal of the method [5.2b]. A population density grating (Sect. 3.4) in a three-level system is assumed [5.2b], as shown in the insert of Fig. 5.6. The grating is produced by absorption of the frequencies ω_1, ω_2 resulting in a spatial and temporal, wavelike modulation of the population density of the energy level E_2. The excited electrons decay from E_2 to the level E_1, thus producing also a modulated population density in the latter level. At frequencies Ω, which are large compared to the reciprocal lifetimes τ_1, τ_2 of the levels, the amplitudes of the population densities decrease because the system integrates rapid fluctuations of the energy density. The modulated population densities of E_1 and E_2 result in a modulation of the optical properties which are detected by diffraction. By suitably choosing the frequency of the probing beam, it is possible to measure separately the population of both levels via diffraction with efficiencies η_1 and η_2. The frequency dependence of η_1 and η_2 determines the lifetimes of E_1 and E_2, as indicated in Fig. 5.6.

5.2 Experiments on Thermal-Energy Excitations

In this section, a number of forced Rayleigh scattering experiments will be reviewed which deal with low energy excitations, namely fluctuations in the temperature and concentration, and in molecular arrangement in fluid systems, population of so-called two-level states in glasses, and convective flow near a hydrodynamic instability. The properties studied are heat flow and second sound, mass diffusion and molecular mobility, glass dynamics, the coupling between some of these modes, and precursors to an instability.

5.2.1 Pure Heat-Flow Studies

The historically first FRS experiments fall into this category. *Eichler* et al. [5.4] created thermal gratings by means of a frequency-doubled Nd:YAG laser. The arrangement is shown in Fig. 5.7a. The samples were either fluids (methanol, glycerol) or ruby crystals. The fluids were made slightly absorbing by the addition of a dye. An argon laser was used for the probe beam and direct detection of the scattered radiation could be employed.

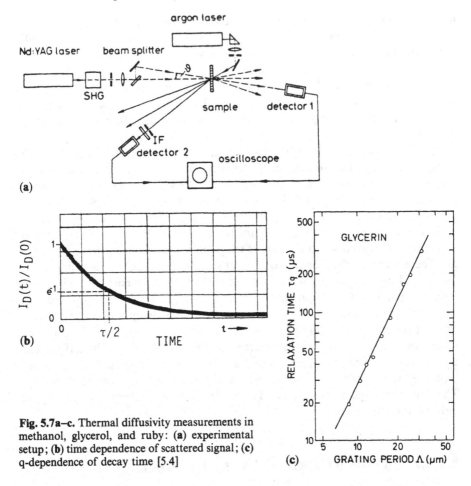

Fig. 5.7a–c. Thermal diffusivity measurements in methanol, glycerol, and ruby: **(a)** experimental setup; **(b)** time dependence of scattered signal; **(c)** q-dependence of decay time [5.4]

Figures 5.7b and c show how well the exponential decay law and the expected q^2 dependence for the decay time are satisfied, compare (3.112, 3.116). Note that the $1/e$ value is already reached at time $\tau/2$, since direct detection monitors the square of the temperature amplitude! From the slope of the decay time vs q^2 curves, the thermal diffusivity was determined, and found to agree well with literature values.

Pohl et al. [5.5] first demonstrated the extremely high sensitivity of FRS by studying heat transport in highly perfect crystals of sodium fluoride (NaF). Interest in this material stemmed from the intriguing finding that at low temperatures heat pulses propagated wave-like rather than diffusive [5.6]. The behavior was considered evidence for the long-predicted phenomenon of second sound in dielectric crystals [5.7]. For a final identification, an independent, different method was highly desirable.

Light scattering was optimal for various reasons, but the scattering rate was too small for classical scattering techniques. This can be estimated from the known thermal expansion, (isobaric) thermooptic coefficient and Pockel's photo-elastic constants of NaF [5.8]. The thermal-expansion coefficient α_l, shown in Fig. 5.8a, illustrates the familiar T^3 temperature dependence for crystalline materials [5.9]. Unfortunately, the thermooptic coefficient $(\partial \chi / \partial T)_{eff}$, calculated on the basis of (3.119–120), decays even faster with decreasing temperature! As a consequence, the expected classical Rayleigh-scattering rate, shown in Fig. 5.8b, drops to undetectably small values upon cooling. The upper curve shows the expected Brillouin-scattering rate for comparison. The shaded area marks the approximate shot noise level at 1 Hz bandwidth.

The corresponding estimate for forced scattering indicates a fair chance for second-sound detection if it existed. In Fig. 5.8c, the solid curve represents the heterodyne scattering rate calculated with the same parameters as before. Note the different slope in the scattering rate vs T as compared to (b), result-ing from the proportionality to $\langle \delta T^2 \rangle$ for classical scattering but δT in FRS (Sects. 2.4.4, 5). The shaded area again indicates the range where a signal-to-noise ratio of the order 1 is expected for 1 s of integration time (1 Hz detection bandwidth). The other details of the figure will be explained below.

Since NaF is highly transparent in the UV, visible and near-IR regions, a pump source in the middle IR was required; CO_2 laser radiation is well suited since $K \simeq 0.4$ cm^{-1} at $\lambda = 10.6$ μm.

The experimental arrangement is shown in Fig. 5.9a. The pump radiation from a 10 W CO_2 laser is sent through the chopper CH and split into two beams, the relative phase, orientation and spot sizes of which are adjusted by means of mirrors M2 to M9. The arrangement allows variation of the very small angle between the pump beams without changing the points of intersection. The latter are in the center of the sample and, for control purposes, at M7. The apparently somewhat involved setup greatly facilitates alignment of the IR beams. The paths of the probe and scattered beams are indicated by dashed lines. A spatial filter behind the sample selects the scattered radiation, which is detected by a photomultiplier (PM) followed by a signal averager.

The response of the scattered signal to a rectangular pump pulse of 1 ms duration is illustrated by displays form the signal averager (Fig. 5.9b). The slow response at room temperature is clearly distinguished from the rectangular shape of the signals at 77 and 20 K. The difference is caused by the strong increase of the thermal conductivity at low temperatures [5.6]. At 20 K, the signal is very weak and overcomes noise after several minutes of integration time only.

The temperature dependence of the experimental data is plotted in Fig. 5.8c (open squares). These data were obtained in the usual manner with a standing grating pattern. They show good agreement with the calculated values (based on diffusion) for $T > 50$ K. A similar temperature dependence of scattering intensity was found by *Eichler* and *Knof* [5.10] for ruby, again showing a many orders of magnitude decrease with decreasing temperature.

The detection of second sound in solids is possible in very high quality crystals within a narrow window of frequencies and temperatures. Experimental verification, though challenging because of setting bounds to the law of heat

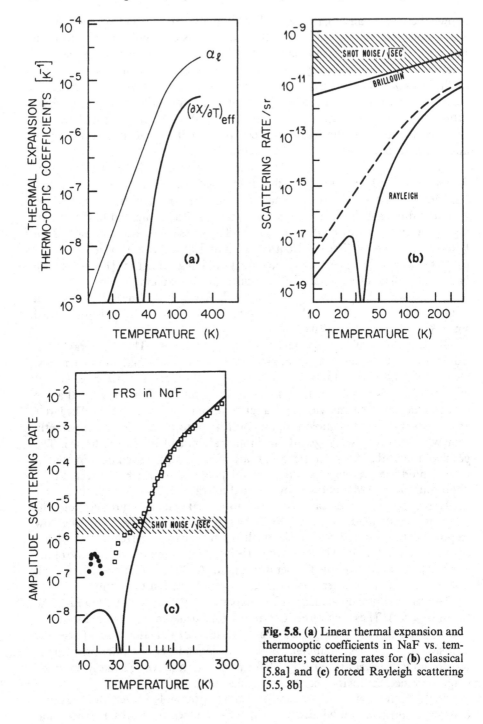

Fig. 5.8. (a) Linear thermal expansion and thermooptic coefficients in NaF vs. temperature; scattering rates for **(b)** classical [5.8a] and **(c)** forced Rayleigh scattering [5.5, 8b]

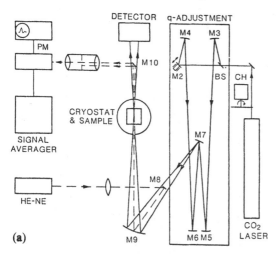

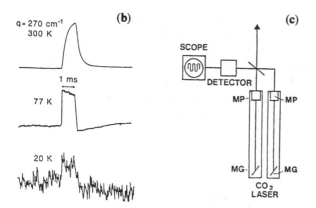

diffusion, is therefore difficult. For detection by FRS, the experimental arrangement described above had to be modified slightly. The pump radiation was amplitude-modulated up to 10 MHz by beating two stabilized CO_2 lasers, as sketched in Fig. 5.9c [5.11]; the FRS signal then carries a component with the same modulation. If second sound really were wavelike, the resonance structure of (3.118) is expected near $\Omega_0 = q/c_0$, c_0 being the velocity of second sound, roughly $1/\sqrt{3}$ times that of regular sound.

A resonance structure was, in fact, found at $\Omega/2\pi \simeq 6.5\,\mathrm{MHz}$ at 20 K, a slightly lower frequency than expected. The solid curves in Fig. 5.10a represent the scattering response at varying temperatures; the dashed line is a control run with blocked pump lasers, and the dash-dotted curve demonstrates heterodyne

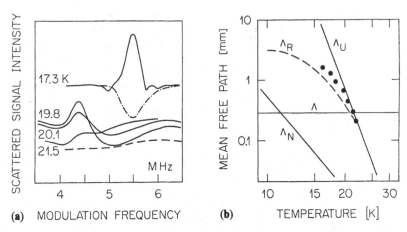

(a)   MODULATION FREQUENCY   **(b)**   TEMPERATURE [K]

Fig. 5.10. (a) FRS heterodyne scattering response to modulated pump radiation in the second-sound regime [5.11a]. (b) Damping length determined from the resonance width together with calculated [5.6] mean free paths [5.11b]

phase reversal. The relation among Ω, q, and c_0 unambiguously identifies the resonance as second sound. The heterodyne scattering rate deduced from the peaks of the structure is shown as solid circles in Fig. 5.8c.

The width of the resonance allowed damping of second sound to be determined for the first time. In Fig. 5.10b, the experimental data are plotted, together with different theoretical damping limits, namely the mean free paths for umklapp (Λ_u), impurity (Λ_R), and normal (Λ_N) processes [5.6, 7]. The wavelength $\Lambda = 2\pi/q$ is also indicated.

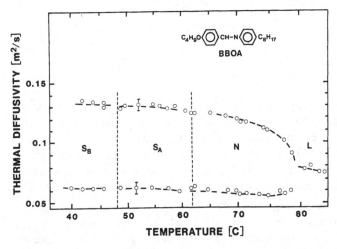

Fig. 5.11. Anisotropic thermal diffusivity in the liquid crystal "BBOA" from FRS [5.12a]. The letter symbols stand for smectic B, smectic A, nematic, and liquid phases, respectively

Urbach et al. [5.12] discuss another interesting application of FRS, the determination of heat conductivities in media which are not easily accessible by classical techniques. The authors investigated the anisotropy of D_{th} in several liquid crystals by FRS. To do this with heaters and temperature sensors would require relatively large samples that cannot be kept easily in a monodomain structure. The FRS technique, however, only needs small volume and small thermal amplitudes, avoiding the risk of heater-induced convection, and the time of measurement is short. A Dove prism in the path of the pump beams allowed rotation of the grating structure with respect to liquid crystal orientation.

Urbach et al. demonstrated the technique on a liquid crystal which exhibits several liquid-crystalline phases, b-butoxy benzylidene p-n-octyl aniline (BBOA). The largest diffusivity was found with q parallel to the molecular axis. This behavior can be followed through the different phases as a function of temperature (Fig. 5.11). Of particular interest are the results for a smectic-C phase showing clearly that the D_{th} anisotropy is caused by the orientation of the molecular axis and not by the layer structure.

5.2.2 Heat Flow Coupled to Other Modes

Anomalous behavior of FRS was found in glasses at low temperatures by *Pohl* [5.14]. Amorphous materials are distinguished from crystals by the possibility of structural rearrangement [5.15]. This process may be described by the motion of a particle in a double-well potential. Fluctuations in the population of these two-level states are expected to influence the low-frequency scattering of light [5.16]. To a large extent, FRS overcomes the signal-to-background problem that restricted the investigation of classical Rayleigh scattering to temperatures above 60 K [5.17].

In Pohl's experiment, optical filter glasses (Schott BG20 and FG18) were chosen as the sample materials. These were mounted on a cool finger for measurements between 25 and 300 K, and inside a pumped He⁴ cryostat for measurements at lower temperatures. Attention was focused on the variations of the thermooptic coupling factor $(\partial \chi / \partial T)_{eff}$ and decay time τ with temperature (Fig. 5.12). At room temperature, $(\partial \chi / \partial T)_{eff}$ is on the order of $10^{-5} \, \text{K}^{-1}$, which is typical for many glasses and crystals. Down to about 70 K, $(\partial \chi / \partial T)_{eff}$ strongly decays with temperature, but below this it becomes practically constant, as compared with the approximate T^3 dependence for simple crystalline materials. The extra scattering appears to be the first-time observation of an influence of rearrangement on the refractive index, as evidenced by its temporal behavior: At sufficiently low temperatures, the decay times show distinct deviations from the curves calculated from heat-conductivity data. They are too large and, most importantly, tend to become independent of the value of q. This is a strong indication for an increasing influence of local relaxation with decreasing temperature. The observed deviation might be caused by a finite relaxation time between different two-level states and the thermal bath of phonons.

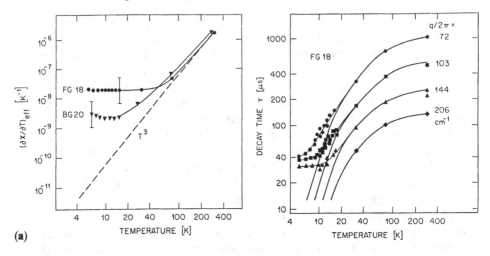

Fig. 5.12a,b. "Anomalous" FRS in glasses at cryogenic temperatures [5.13]: thermooptic coefficient (**a**) and decay time (**b**)

Cowen et al. [5.17] demonstrated the interaction between a temperature grating and the so-called Mountain mode [5.18] in liquids. Two characteristic times could be distinguished within a certain temperature window (Fig. 5.13a). The data, taken in heterodyne configuration, show a rapid positive decay for $T = 60.5\,°C$ which represents the response of forced Mountain scattering. Note that the rearrangement of molecules responsible for the Mountain mode is local relaxation; the fast exponential is therefore independent of $|q|$.

Chan and *Pershan* [5.19] measured the diffusion of partially bound water between smectic lipid membrane layers by FRS (Fig. 5.13b). In the plot, the two decay times have already been extracted from the raw data. The graph demonstrates the diffusive character of both decay processes. In their investigation, the decay time of the overall heating (finite spot-size effect) was quite close to the membrane diffusion time but could be discriminated by periodic inversion of both phase and signal.

A Benard cell is a flat container completely filled with a liquid that can be heated from below and cooled from the top. At a certain temperature gradient, convection sets in to form a regular flow pattern. This so-called Benard instability has many properties in common with a phase transition [5.20]. The question was whether temperature fluctuations coupled to velocity would critically slow down close to the transition. *Allain* et al. [5.21] demonstrated this phenomenon in an FRS experiment by adjusting q to the anticipated flow-pattern structure and watching the decay time upon approach to the critical gradient (Fig. 5.13c). They used the arrangement of Fig. 2.11h to achieve the large wavelengths Λ indicated.

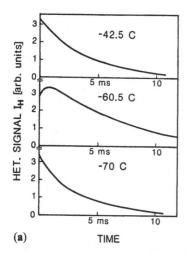

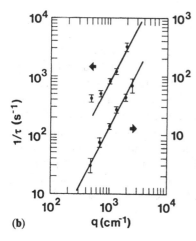

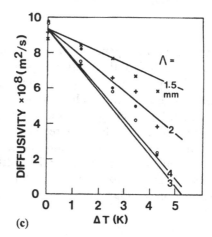

Fig. 5.13a–c. Various examples of temperature-coupled modes detected by FRS: (a) Mountain plus temperature modes in glycerol at 60.5 °C [5.17]; (b) the two decay times observed in smectic lipid-membrane solutions [5.19]; (c) temperature/velocity-coupled diffusivity close to the Benard instability [5.21]

Boon et al. [5.22] demonstrated the existence of propagating heat waves in an inverted Benard cell, i.e. one which is heated from above such that the thermal gradient tends to suppress convective fluctuations.

The grating amplitudes in all the experiments discussed up to now were small compared to the average sample temperature. This obviously is a prerequisite for well-defined, easy-to-interpret scattering dynamics. Large amplitudes, on the other hand, allow to create interesting new experimental conditions such as the transient-grating study under annealing conditions by *Marine* et al. [5.23].

The standard annealing experiment employs a simple circular laser spot for heating. By replacing the uniform spot with an interference pattern, Marine et al. were able to detect self-diffraction and forced scattering during and after excitation with a 40 ns ruby laser pulse. Thus, they got quantitative information on the (complex) refractive index of the GaAs film during laser heating.

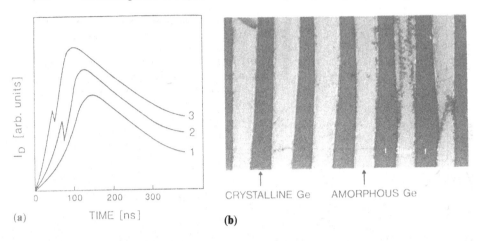

Fig. 5.14a,b. Recrystallization of a-GaAs and Ge observed (**a**) by FRS [5.23] and (**b**) in an electron microscope (photograph courtesy of W. Marine)

The recorded time dependence of the scattered signals from a cw argon laser is shown in Fig. 5.14a for three different pump energies. A number of interesting features can be recognized: (1) the maximum scattering rate is reached well after the end of the pulse, which may have to do with thermalization delay of excited electrons; (2) at higher energies, a sharp kink is seen at time t_2 which has been attributed to melting; (3) the scattered signal does not return to zero, since the recrystallization process at the pump interference maxima is permanent. The height of the plateau is a measure for the width of the recrystallized stripes and their degree of recrystallization. Figure 5.14b depicts a Ge film surface after grating formation. Similar effects are described in Sects. 3.5.5b and 5.9.

5.2.3 Concentration Gratings and Mass Diffusion

We devote a subsection of its own to mass diffusion experiments although most of them also belong into the category of coupled modes.

As pointed out in Sect. 3.7.4, a thermal grating in a liquid mixture tends to create a certain amount of phase separation, i.e. a concentration grating (the Soret effect). *Thyagarajan* and *Lallemand* [5.24] demonstrated this phenomenon using a mixture of carbon-disulfide and ethanol. The refractive indices of the two components are sufficiently different to provide a large coupling coefficient $\partial \chi / \partial \tilde{c}$. Pumping with a 2 W argon laser and detecting with an He-Ne laser in heterodyne configuration, those authors could simultaneously see both thermal and concentration gratings, decaying with vastly different time constants τ_{th} and τ_m, respectively (Fig. 5.15a). From these data, essentially the amplitude ratio of the two exponentials, the Soret constant k_T has been determined for various concentrations of this mixture (Table 5.1). This is a convenient, fast, and non-contact method as compared to classical techniques using external heaters, heat sinks, thermometers, and composition analysis.

The Soret effect was also used by *Pohl* [5.25] to study the dynamics of the concentration mode near a critical consolute point (c_c, T_c) where the (mutual) mass diffusion becomes soft and, after going through T_c, divergent (Sect. 3.7.3).

In this experiment, a critical mixture of 2,6-lutidine/water (LW) was used which has an inverted curve of coexistence. (Such a mixture phase-separates upon heating, while most mixtures do this upon cooling). The critical point is at 31 % vol./32 C [5.25]. Because of its inverted characteristics, LW can be quenched by heating, which is more readily at the experimenter's disposition than cooling.

The LW mixture was temperature-stabilized to within 1 mK. By employing pump pulses from an argon laser, the average temperature at the center could be raised by up to 100 mK while the thermal grating was created. The mixture was slightly colored with methyl red for this purpose. The temporal development of the center average temperature, calculated from (3.115) is shown for different starting temperatures and pulse strengths in Fig. 5.15b. The change in $T_{av} - T_c$ is relatively small for trace A. Trace B brings T_{av} into the immediate vicinity of T_c for a very short time. Trace C indicates a very short quench above T_c, while in D the system dives deeply into the forbidden zone above T_c.

The temporal evolution of the respective heterodyne FRS signals is shown in Fig. 5.15c. A is a well-defined exponential indicating negligible influence of the T_{av} excursion; B is similar but slower; C shows a slight increase after the end of the pulse and an extremely slow decay (note the time scale!). The most interesting curve, however, is D: Here, a growth of scattering can be seen which extends for more than 2 seconds (while the pump ended after 0.4 s); the scattered signal then stays almost constant for another 2 seconds, before it finally and slowly begins to decay.

From these data, it is possible to extract an instantaneous growth (decay) rate (Fig. 5.15d), $(1/\tau)_{inst} \equiv \partial I_H / \Delta I_H \partial t$, where $\Delta I_H \equiv I_H(t) - I_H(t \to \infty)$. The negative part – stemming from the stable side of the phase diagram – is in good agreement with previous classical light scattering data [5.27]; the zero transition and positive part – stemming from the dive into the forbidden regime – represents

Table 5.1. Decay times for a 19 μm wavelength grating in carbon disulfide-ethanol mixtures, amplitude ratio and Soret constants [5.24]

Concentration [mass CS$_2$/total mass]	τ_{th} [μs]	τ_m [ms]	Amplitude ratio	k_T
0.24	122	6.0	450	0.34
0.324	129	6.0	335	0.40
0.398	147	5.5	292	0.29
	126	4.4	320	0.30
0.44	132	4.0	226	0.34

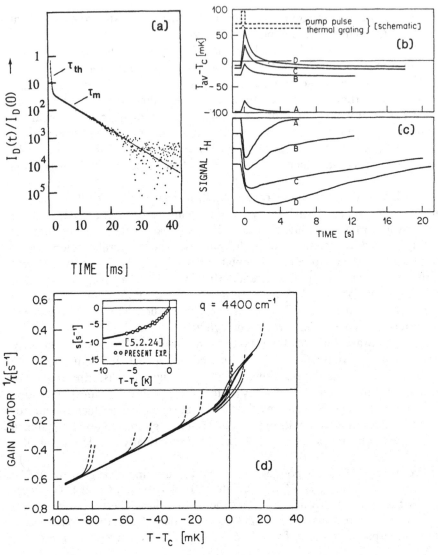

Fig. 5.15a–d. Mass diffusion in binary liquid mixtures: (**a**) FRS signal showing both the fast decay and the slow concentration decay [5.24]; (**b**) average temperature and (**c**) FRS signal near a critical consolute point [5.25]; (**d**) growth rate $1/\tau$ resulting from **b** and **c**, inset: comparison with corresponding classical scattering experiment [5.27]

entirely new information accessible with no other means than FRS so far. The different curvelets represent different starting conditions and pulse power; attention should be focused on the common envelope (thick trace).

A different way for producing a grating was chosen by *Hervet* et al. [5.28]. Photochromic dye molecules were introduced to the organic solutions (mostly polymers and liquid crystals) they were interested in. The pump pulses created a

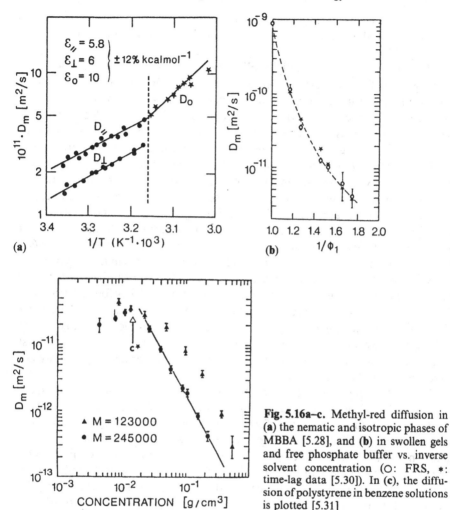

Fig. 5.16a–c. Methyl-red diffusion in (a) the nematic and isotropic phases of MBBA [5.28], and (b) in swollen gels and free phosphate buffer vs. inverse solvent concentration (O: FRS, *: time-lag data [5.30]). In (c), the diffusion of polystyrene in benzene solutions is plotted [5.31]

periodic variation of photoexcited molecules which primarily acted as an absorption grating. The strong diffraction from the grating allowed direct detection.

The lifetime of the excited state of the dye used – methyl red – is of the order of several seconds. For sufficiently large q, grating decay by diffusion occurs within a fraction of a second, and hence is a measure for the diffusity of these tracer molecules in the solution under investigation.

Figure 5.16 shows the temperature dependence of the mass diffusion coefficient in the liquid crystal MBBA obtained in this way. The single curve on the right, representing the isotropic low-temperature phase of this material, finds its continuation in the diffusivity of the nematic phase parallel to its optic axis (left-hand part of the graph). Diffusion perpendicular to this axis, on the other hand, is considerably reduced. This result is in excellent agreement with previous data which, however, were more bothersome and time consuming to obtain [5.29].

Wesson et al. [5.30] recently applied this dye tracer/FRS technique to the investigation of diffusion in swollen gels. They carefully compared the data with those obtained with the classical time-lag method. The results indicate structural homogeneity of the gels down to the micrometer size, and perfect agreement between the two methods. This is demonstrated in Fig. 5.16b where values for the diffusivities for gel and sol states obtained by the two methods are shown.

The dynamics of entangled-polymer-chain solutions is difficult to approach experimentally with conventional methods, since it involves very slow diffusion. *Hervet* et al. [5.31] were able to measure the diffusion of such molecules (polystyrene) systematically for the first time by using FRS and relating it to scaling and reptation models. As a function of concentration, a sharp crossover was found between dilute and entangled concentration regimes (Fig. 5.16c). Note the extremely small values of the diffusion coefficients in this experiment!

Recently, *Nemoto* et al. extended the range of diffusion studies even further by exploring self-difusion in amorphous polymers [5.32]. They found FRS decay times of the order of hours corresponding to diffusion coefficients of the order $10^{-16} \, \text{m}^2/\text{s}$! Because of this high resolution, it was possible to compare mass diffusion and shear viscosity within a sizeable temperature range for the first time. The results yield information on the chain dynamics of the polymer molecules with the free-volume theory.

The method of photochromic labelling was combined with electrophoresis as an alternative to electrophoretic light scattering by *Johnson* and co-workers [5.33], and *Yu* and co-workers [5.34]. The labelled species migrate along the electric-field lines which are arranged parallel to q. This results in an increasing phase shift in heterodyne detection, i.e., ondulations in the signal received. Their frequency provides electrophoretic mobilities, relative phase shifts, and relative numbers of labels in each component of the mixture. In Yu's experiment, the sample cell is split into two halves: The front part of the cell is equipped with electrodes and provides the electrophoretic signal, while the rear part – identical in dimensions with the front part – is field-free and provides a regular FRS signal which acts as reference. These experiments bear close similarity to the flow studies discussed in Sect. 5.4.

The current work on polymer diffusion by FRS has recently been reviewed by *Yu* and co-workers [5.35].

5.3 Laser-Induced Ultrasonics

The dynamic-grating experimental arrangement has been used for the optical generation of ultrasonic waves. Optical excitation and probing results in a very versatile method for sound-wave investigations.

In the first experiments [5.36, 37], two laser beams with slightly different frequencies were used to produce sound waves in liquids and solids by electrostrictive mixing. The two light waves produce a moving interference pattern which resonantly excites a sound wave if the frequencies and velocities are matched. These experiments are strongly related to stimulated Brillouin scattering where only one incident wave is present and a second wave builds up from noise, as outlined in Sect. 1.2.2.

In later work [5.38–50], sound-wave excitation was achieved with short laser pulses and spatially periodic illumination. A schematic illustration of a typical experiment is given in Fig. 5.17. Two single picosecond pulses produce an interference pattern with a grating spacing Λ of 0.1 to 1000 µm. The pulse width should be short compared to the acoustic oscillation period to obtain efficient sound excitation. Due to absorption, the material is heated and thermal expansion results in a spatially periodic density change. In transparent or weakly absorbing materials, density changes due to electrostriction or the photoelastic effect are also important. Because of the elastic response of the material, a standing density wave is excited. The sound-wave frequency is given by $f = v/\Lambda$. For liquids, the sound velocity v is about 1000 m/s and frequencies (f) of 1 MHz to 10 GHz are obtained. For solids, the sound velocity is higher, and frequencies up to 50 GHz are possible.

Various kinds of ultrasonic waves can be excited. Experimental work has mainly been done on surface (Rayleigh) waves on fused silica, quartz and metals [5.38, 40], and on longitudinal waves, in methanol [5.41], ethanol [5.47], other solvents [5.49] and the organic crystals p-terphenyl [5.45, 50] and perylene [5.46].

The production of longitudinal sound waves (density waves) in liquids by a thermal coupling mechanisms was described by the linearized hydrodynamic equations [5.45], which are used also for the mathematical formulation of stimulated thermal Brillouin scattering, as given by

$$\frac{\partial^2 \varrho}{\partial t^2} + v^2 \nabla^2 \varrho - \frac{\eta}{\varrho_0} \frac{\partial}{\partial t} \nabla^2 \varrho = v^2 \beta \varrho_0 \nabla^2 T , \tag{5.1}$$

$$\frac{\partial T}{\partial t} - \frac{\lambda}{\varrho_0 c} \nabla^2 T = \frac{\gamma - 1}{\beta \varrho_0} \frac{\partial \varrho}{\partial t} + \frac{K}{\varrho_0 c} I . \tag{5.2}$$

Equation (5.1) is the acoustic-wave equation for the density ϱ with a driving term due to the temperature T, while v is the adiabatic sound velocity, η the viscosity leading to damping, β the thermal expansion coefficient, ϱ_0 the average density, and $\gamma = c_p/c_v$ the ratio of the specific heats. Equation (5.2) is the heat-conduction equation with heat-production terms due to heating by the light intensity I and compression of the material, where λ is the thermal conductivity, c the specific heat, and K the absorption coefficient. The optical response is given by

$$\Delta n = \frac{\partial n}{\partial \varrho} (\varrho - \varrho_0) .$$

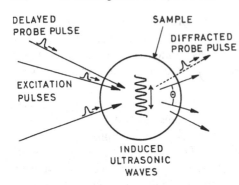

Fig. 5.17. Experimental schematic for all optical sound wave excitation and detection using the grating arrangement. After *Nelson, Miller, Fayer* et al. [5.46–50]

In solids transverse and mixed sound-waves are also possible and it is necessary to describe thermal expansion and the subsequent sound dynamics by calculating the stresses and strains. In general, one quasi-longitudinal and two quasi-transverse waves will be generated with wave vector equal to that of the grating. Along pure mode directions, pure transverse waves will be excluded.

It has been pointed out [5.48] that transverse or shear waves can also be excited if the polarization of the two excitation pulses are not parallel but optimized for the generation of a sound wave (phonon) of a particular polarization. The photoelastic coupling mechanism has to be used in this case and the acoustic response to an applied electromagnetic field is given in the cgs system by [5.48]

$$\varrho_0 \frac{\partial^2 u_i}{\partial t^2} - \Sigma_{j,k,l} \lambda_{ijkl} \frac{\partial^2 u_i}{\partial x_j \partial x_k} = \Sigma_{j,k,l} \frac{1}{8\pi} P_{ijkl} \frac{\partial}{\partial x_j} D_k D_l \; . \tag{5.3}$$

Here the u_i are the material displacements, λ_{ijkl} the elastic stiffness constants, p_{ijkl} the photoelastic constants and D_i the electric displacements inside the sample. The strains are obtained from the material displacements by differentiation, and the optical response can be calculated by using the elastooptical parameters of the material.

Sound waves have been excited not only by single-pulse pairs, but also by double-pulse pairs [5.47] and pulse sequences [5.42]. If the time delay between the two pulses matches the oscillation period of the sound wave, the amplitude is resonantly enhanced. It is thus possible to use consecutive pulses from a mode-locked laser to obtain a large acoustic amplitude.

The optically excited sound waves can be detected by diffraction of a probe pulse, as indicated in Fig. 5.17. The diffraction efficiency as a function of the delay-time is a measure for the sound-wave amplitude. Often, not only the acoustic grating is present but also a slowly varying thermal or population density contribution. If a cw probe is used [5.41, 42] the total time dependence of

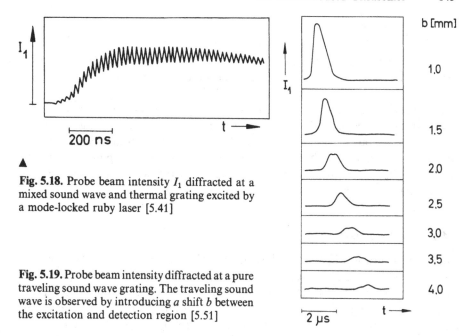

Fig. 5.18. Probe beam intensity I_1 diffracted at a mixed sound wave and thermal grating excited by a mode-locked ruby laser [5.41]

Fig. 5.19. Probe beam intensity diffracted at a pure traveling sound wave grating. The traveling sound wave is observed by introducing *a* shift *b* between the excitation and detection region [5.51]

the grating can be observed. An example is given in Fig. 5.18. The sound wave was produced by a mode-locked ruby laser with a pulse width of about 200 ns containing about 10 mode-locked pulses. The mode-locking frequency of about 50 MHz was adjusted to be equal to the sound frequency so that the consecutive mode-locked picosecond pulses build up a large sound amplitude. A quasi cw argon laser beam was diffracted at the sound wave and a thermal grating was built up simultaneously. As shown in Fig. 5.18, the sound-wave amplitude decays rapidly compared to the thermal grating amplitude.

The standing sound wave produced in the interference region of the excitation laser beams decays into two moving sound waves. Propagating sound waves can thus be detected independently of the temperature grating which keeps its position in the excitation area. For this purpose, the probing argon laser beam was shifted. If the detection and excitation regions are completely separated, an unmodulated diffracted pulse appears with a delay corresponding to the time in which the sound wave packet travels from the excitation to the detection region (Fig. 5.19). The displacement allows the determination of the sound velocity, while the amplitude decrease gives the sound damping.

Compared to conventional ultrasonic methods, the grating technique has the advantage that sound waves can be excited in microscopically small areas without external transducers. It is therefore possible to investigate small samples or to scan the ultrasonic properties of inhomogeneous materials. Besides measuring acoustic parameters, it has been demonstrated [5.49] that the method is also useful to determine weak absorption and elastooptic constants.

5.4 Flow Studies

Forced Rayleigh scattering has been applied to study the velocity and velocity gradients in liquid flow [5.52–55]. In addition, laser-induced thermal gratings have been used for flow visualization [5.54]. The experimental technique starts with the generation of a localized thermal grating inside the flowing liquid (e.g., water or methanol with an absorbing dye) by a pulsed laser (e.g., a Nd:YAG laser with $\lambda = 1.06 \, \mu m$; 80 mJ in 15 ns). The pulse width should be short enough so that the movement of the liquid can be neglected during the pulse. Due to the flow of the liquid, the grating moves or becomes distorted; this can be observed by diffraction of a probe beam or by imaging the grating.

5.4.1 Flow Visualization

The thermal grating in flowing liquids is probed with a delayed laser pulse. Experimentally, an acousto-optically switched He-Ne laser has been used. The two first-order diffracted beams are superimposed with a lens, resulting in an image of the thermal grating. The phase modulation of the grating is transformed into an amplitude modulation in the image plane, as usual in phase-contrast or strioscopic techniques, e.g., by slight defocusing. 100 % contrast is obtained by blocking the central beam.

By varying the time delay between the excitation and probe pulses, the distortion of the grating with increasing time can be followed. As an example, the parabolic velocity profile in a laminar Poiseuille flow was demonstrated with this technique by *Fermigier* [5.54].

5.4.2 Flow Velocity

If the grating is probed with a sufficiently long laser pulse, the fringes in the image of the grating move with the flow velocity because the object (grating planes in liquid) also moves. A more sophisticated way to understand the fringe movement in the image is to consider the frequency shifts $\pm \Delta v$ in the diffracted beams due to the movement of the grating (Sect. 4.7). Superposition of these two beams with slightly different frequencies results in a moving interference pattern.

The movement of the grating image is detected with a small-area photodetector which sees a temporal intensity modulation with a frequency $2\Delta v = \Lambda/v$, where Λ is the grating period, and v is the flow velocity averaged over the grating period.

The technique described above is complimentary to the classical laser Doppler anemometry where the frequency shift of light scattered at tracer particles moving with the flow is detected [5.56]. The grating technique offers the advantages that (1) it does not require the presence of tracer particles, (2) it allows also the measurements of other properties of the flow, such as heat or mass diffusivities, which are important in turbulent flow.

An alternative method for the measurement of flow velocities consists in observing the time-averaged amplitude of a grating excited in a flow. It is assumed now that the grating is not excited by a single short pulse, but by a sequences of pulses or by a long pulse with a width larger than Λ/v. If the grating planes are perpendicular to the flow, the grating amplitude is strongly dependent on the flow velocity which can wash out the grating [5.57, 58]. If the grating planes are parallel to the flow, the grating amplitude is only weakly affected by the flow, if the size of the grating is sufficiently large. It is therefore possible to determine the flow velocity by measuring the ratio of the grating amplitudes or the diffraction efficiencies with the grating planes parallel and perpendicular to the flow velocity. Fast photodetectors are not needed for this purpose. The technique may be useful for optical laboratories using liquid flows only as tools, e.g. for the determination of flow velocities in jets for dye lasers.

5.4.3 Velocity Gradients

A grating with an initial grating-vector $q_i(0)$ is transformed by an inhomogeneous flow after a short time t into a grating with a final grating-vector $q_i(t)$ given by

$$q_i(t) = q_i(0) - \Sigma_j q_j(0) s_{ji} t \quad \text{if} \quad s_{ji} t \ll 1 \; , \tag{5.4}$$

where $s_{ji} = \partial v_i/\partial x_i$ is the velocity gradient tensor [5.52, 55]. The transformation can lead to a rotation of the grating or a change of the grating period. A simple example for a rotation is given by a Poiseuille flow with the velocity increasing from the walls to the center of the cell (Fig. 5.20). An example for a change of the grating period is provided by a laminar flow through a tube with changing cross-section where the velocity changes in the direction of the flow.

The grating rotation or period change can be detected by comparing the diffraction pattern of two light pulses with delay time t. Experiments demonstrating the method have been carried out in a laminar plane Poiseuille flow giving the expected linear dependence of the velocity gradient $\partial v_1/\partial v_2$ on the distance x_2 from the center of the flow cell.

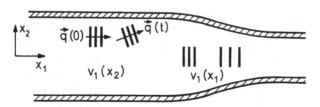

Fig. 5.20. Grating rotation or straining in a flow according to (5.4). In a shear flow with velocity $v_1(x_2)$, a grating vector $\mathbf{q}(0) = (q, 0, 0)$ is transformed into $\mathbf{q}(t) = (q, -qt\partial v_1/\partial x_2, 0$. Note that $(\partial v_1/\partial x_2)$ is negative for $x_2 > 0$. The rotation angle is given by $t\partial v_1/\partial x_2$. In a flow with a longitudinal gradient, the final grating vector is given by $(q - qt\partial v_1/\partial x_2, 0, 0)$ which corresponds to an increase of the grating period

The method is also useful for the investigation of turbulent flow where a statistical description is necessary in terms of Richardson functions giving the progressive separation of two particles [5.52, 55].

5.5 Investigation of Carrier Dynamics in Semiconductors

Laser-induced free-carrier gratings (Sect. 3.5) decay by a number of different physical processes which can be investigated by time-dependent observations of diffracted probe beams. Special grating decay processes are recombination and diffusion. Recombination occurs also in a homogeneously excited sample and can be probed simply via the absorption change due to carrier excitation [5.59, 60]. In contrast, diffusion takes places only if a spatial modulation of the carrier density is present. Therefore laser-induced grating experiments are especially useful to investigate diffusion processes.

The temporal and spatial dependence of the optically excited carrier density $N(x, t)$ is described by the one-dimensional diffusion equation

$$\frac{\partial N(x, t)}{\partial t} + \frac{N(x, t)}{\tau} - D_a \frac{\partial^2 N(x, t)}{\partial x^2} = \frac{\zeta K I(t)}{h\nu} [1 + \cos(2\pi x/\Lambda)] \ . \tag{5.5}$$

For simplicity, it is assumed here that the carrier density depends only on the coordinate x in the grating direction. This implies that the sample thickness is large compared to the grating period Λ. In addition, homogeneous excitation is assumed, so that the spatial amplitude $I(t)$ of the light intensity distribution does not change along the thickness of the sample, i.e., the absorption coefficient K has to be sufficiently small. The other symbols in (5.5) are the recombination time τ of the electron-hole pairs, the ambipolar diffusion constant D_a, the photon energy $h\nu$ and, the quantum efficiency ζ for the excitation of an electron hole-pair by an incident photon. At low excitation, the quantum efficiency ζ is often assumed to be 100%. At higher excitation, various free-carrier absorption processes have to be taken into account and the quantum efficiency drops [5.61]. For the following simplified discussion $\zeta = 1$ is assumed.

Equation (5.5) implies that the optically excited free carriers, i.e. electron and holes, diffuse so that their densities are equal and no space charge develops. This is called ambipolar diffusion, which denotes only a simplified description. In general, diffusion has to be considered for the electrons and holes separately, with individual diffusion constants. Also, the space-charge and the coupling of the electron and hole densities due to the electric space-charge field has to be taken into account. In a simplified picture, one can describe the electric attraction of the electron and hole distribution by electron-hole pairs bound by the Coulomb force and diffusing together, so that (5.5) is obtained. It should be possible, however, to measure also the individual diffusion of electron or holes by using strongly doped n-type or p-type samples.

Another approximation in (5.5) concerns the recombination lifetime τ. In semiconductors, the recombination rate N/τ is strongly carrier dependent, which may be approximated by a Taylor expansion

$$N/\tau = AN + BN^2 + CN^3 + \ldots . \tag{5.6}$$

The coefficients A, B, C in this equation can be attributed, at least approximately, to different physical mechanisms. Linear recombination may be due to impurities and some background carrier concentration. The coefficient B describes bimolecular recombination, which is mostly due to radiative recombination. C comes from Auger processes where the energy of a recombining electron and hole is transfered to another carrier which is excited to a higher energy in the valence or conduction band. With the high optical excitation used for grating experiments, the nonlinear recombination processes dominate, so that τ is not constant. Equation (5.5) then becomes strongly nonlinear and difficult to solve. To avoid this complication one may restrict the solution of (5.5) to a small regime of the carrier density, so that τ can be approximated by a constant value. In addition, most experiments on laser-induced free-carrier gratings have been carried out under condition where diffusional decay dominates, so that recombination has a comparatively small influence on the grating decay time.

The solution of (5.5) with arbitrary excitation pulse shape $I(t)$ was given in [5.61]. For the simple case of excitation with a temporal Dirac pulse one obtains

$$N(x, t) = N_0 [1 + \exp(-t/\tau_D) \cos(2\pi x/\Lambda)] \exp(-t/\tau) . \tag{5.7}$$

The diffusive decay time τ_D is given by

$$\tau_D = \Lambda^2/4\pi^2 D . \tag{5.8}$$

The spatial amplitude $N(x, t)$, i.e. the grating amplitude, decays with the grating decay time τ_g:

$$1/\tau_g = (1/\tau_D) + (1/\tau) = 4\pi^2 D/\Lambda^2 + (1/\tau) . \tag{5.9}$$

The grating decay time τ_g is measured by observation of the time dependence of a diffracted probe beam. With weak excitation, the diffracted intensity I_1 depends quadratically on the complex refractive index change $\Delta \tilde{n}$ and on the carrier density N, i.e. $I_1 \sim |\Delta \tilde{n}|^2 \sim (N)^2$. Therefore, the decay time τ' observed in the diffracted intensity amounts to half the grating decay time, i.e. $\tau' = \tau_g/2$. At high excitation, the diffracted intensity depends in a more complicated form on the refractive index change, e.g. in case of a thin, strong phase grating the dependence is given by a squared Bessel function (Sect. 4.2.2). The evaluation of grating decay times τ_g then becomes more complicated [5.61].

The relative contribution of diffusion and recombination to the grating decay time τ_g can be controlled according to (5.9) by choice of the grating spacing Λ.

If typical values of $D_a = 10\,\text{cm}^2/\text{s}$ and $\tau = 10^{-8}\,\text{s}$ are assumed, it is necessary to choose $\Lambda < 20\,\mu\text{m}$ in order to have diffusional decay dominant.

Ambipolar diffusion constants have been measured in a number of semiconductors, e.g. Si [5.62, 63], Ge [5.64], CdS, CdSe and ZnSe [5.65]. In Si and Ge, the results agree well with values calculated from mobility data, whereas in CdS the nonlinear transient grating method gives a much larger hole diffusion coefficient than obtained in other experiments[5.66]. Recombination and diffusion in amorphous Si have been studied in [5.104–106], hot carrier diffusion in [5.107].

Diffusion measurements have been carried out not only using short-pulse excitation and observing the grating decay, but also with excitation pulses that were long compared to the grating decay time. In the latter case, the grating amplitude is measured as a function of the grating spacing Λ, i.e. as a function of the grating decay time. In this case, τ_g determines the steady-state carrier density and grating amplitude and may then be evaluated without a time-resolved measurement.

The grating technique has been applied also to investigate surface recombination in semiconductors. The optically excited carriers may recombine also at the surface of the sample, in addition to the recombination process in the bulk discussed up to now. Surface recombination is especially important when the excitation radiation is strongly absorbed, so that the carriers are created only in a thin layer near the surface. The recombination is then described by a surface recombination velocity and an appropriate boundary condition for the solution of (5.5). It appears that the carrier decay in a thin surface layer depends on the recombination velocity and also on the layer thickness and ambipolar diffusion constant which controls the transport of the carriers to the surface. A detailed theoretical and experimental discussion of surface recombination in gratings experiments is given in [5.67, 68].

Further studies of carrier dynamics have been performed in Germanium [3.80–90] and InSb [3.189], as mentioned already in Sect. 3.5 where material investigations are summarized.

5.6 Electronic Energy (Exciton) Transfer in Solids

Optically excited centers in solids or other materials may transfer their energy to other centers. This transfer of optical excitation energy is important in a number of processes in nature and technology, such as photochemistry, photosynthesis and solar-energy collection. Another example is solid-state-laser technology. In a laser, the excited-state population is depleted inhomogeneously and the question is whether the spatial population modulation is washed out by diffusion. The effect of this spatial hole burning effect on laser performance is discussed in Sect. 7.1.

A large number of theoretical and experimental investigations on energy transfer has been performed, and we give here only some arbitrarily selected references on work in inorganic crystals [5.69] and dye solutions [5.70, 71].

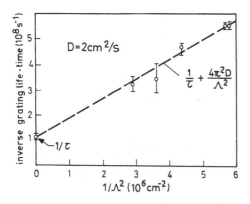

Fig. 5.21. Energy transfer in pentacene doped into p-terphenyl; The graph shows experimental grating decay time vs. grating period Λ after *Salcedo* et al. [5.78]

Forced light scattering at a laser-induced grating can be used to investigate energy diffusion in a similar way, as described in Sect. 5.5 for the measurement of carrier diffusion in semiconductors. The method consists essentially of measuring the decay time of a spatially modulated distribution of the excited states to be investigated. The exponential grating decay time τ_g is given by the recombination life-time τ of an excited center and the energy diffusion constant D:

$$1/\tau_g = 1/\tau + D/4\pi^2\Lambda^2 \; . \tag{5.10}$$

The grating decay time τ_g is usually measured as a function of the grating period Λ. A plot of $1/\tau_g$ versus Λ^2 gives the recombination time τ and the diffusion constant D, as indicated in Fig. 5.21, which has been obtained for the organic crystal p-terphenyl doped with pentacene [5.78].

The grating technique is useful to investigate energy diffusion processes if these processes contribute sufficiently to the grating decay. From (5.10), the following condition for the grating period is obtained

$$\Lambda \leq F\sqrt{D\tau} \; .$$

Here $\sqrt{D\tau}$ is called the diffusion length. F is a factor of the order of 1, which depends on the accuracy of the measurement. The smallest grating periods that can be produced are given by $\Lambda = \lambda/2n$, where λ is the excitation wavelength and n is the refractive index of the material. Using visible light ($\lambda = 500$ nm) and highly refractive materials ($n = 2.5$), the grating period is $\Lambda = 100$ nm $= 1000$ Å and energy diffusion over this distance is detectable. On the other hand, it seems difficult to measure energy diffusion with the grating technique if the diffusion length amounts only to a few atomic distances, as is expected for many materials.

One of the most frequently studied inorganic systems is ruby, Al_2O_3 doped with Cr^{3+} as active centers. The reason is that the Cr^{3+} ions have a fairly well understood energy level scheme, and many other physical properties are also very well known. Also, ruby was the first material in which laser action has been

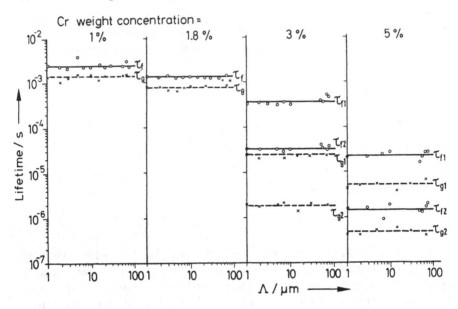

Fig. 5.22. Absence of energy transfer in ruby; The graph shows experimental grating decay time τ_g vs. grating period Λ for various Cr^{3+} concentrations. For comparison, the fluorescence decay time τ_f is also given. At high concentrations, double exponential decay is observed [5.74] so that two decay times are given

accomplished and therefore has been of great technological interest. Additional interest for ruby systems occurs because ruby has been considered as a candidate to observe a transition between low mobility of the excited energy state to high mobility above a critical density of the excited centers [5.72]. Such an *Anderson* transition [5.73] is also called a mobility edge. At low density, the centers are localized, whereas delocalization is expected at high density. Similarly, the electronic wave functions in a periodic crystal potential are spread all over the crystal.

The grating technique has been applied to measure energy transfer between single Cr^{3+} ions, whereas the energy transfer from single ions to Cr^{3+} pairs having a different spectral emission is detected with fluorescence techniques.

The grating or four-wave-mixing experiments (Fig. 5.22) performed by various authors [5.74–76] with Cr^{3+} concentration from 0.05 to 5 % have shown that the grating decay time τ_g is independent of the grating period Λ, i.e. diffusion of the excited state energy has not been detectable.

Spatial energy transfer has been investigated also in $Nd_xLa_{1-x}P_5O_{14}$ crystals. These materials have attracted interest as minilasers for low-threshold, high-gain applications. The presence of energy migration has been observed with the grating technique [5.77].

5.7 Charge Transport in Electrooptic Materials

Laser-induced gratings can be used for studying charge-transport processes in photorefractive materials. The phase gratings in these materials are produced by photo-induced space-charge fields which give rise to refractive index changes by the electro-optic effect (Sect. 3.6). The space-charge fields are induced by transport mechanisms such as diffusion, photo- and dark conductivity, as well as the photovoltaic effect. Electrons, holes and ions can, in principle, be the charge carriers. The laser-induced grating technique allows one to study the relative contributions of each transport mechanism and the determination of the sign of the major charge carriers.

The grating erasure times are given by the Maxwell time constant (3.101)

$$T_0 = \frac{\varepsilon\varepsilon_0}{\sigma_0 + bKI} \, , \tag{5.11}$$

where ε is the dielectric constant, $\sigma_0 = e\mu_e n_d$ is the dark conductivity, KI the absorbed intensity and $b = \zeta\mu_e\tau e/hv$ is a constant which describes the photoconductivity (Sect. 3.6.5). (μ_e: electron (hole) mobility, n_d: thermally excited electron (hole) concentration, ζ: quantum efficiency, τ: carrier lifetime and hv: photon energy.) A measurement of the intensity dependence of the grating build-up time thus allows one to determine without mechanical contact the dark conductivity or the thermally excited dark-carrier concentration n_d and the photoconductivity bKI or the product $\zeta\mu\tau$.

An example of hologram writing and erasure in $KNbO_3$: Fe is shown in Fig. 5.23. As can be seen in this figure, the writing is much faster than erasure in the dark, because photoconductivity contributes in the former case.

The relative contributions of the above-mentioned charge transport mechanisms can also be studied by the grating technique, because the relative influence of diffusion increases with decreasing fringe spacing [increasing spatial gradient of photoinduced carrier concentration, see (3.82, 3)], the drift contri-

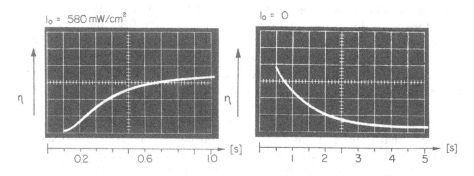

Fig. 5.23. Photorefractive grating write-erase cycles in a $KNbO_3$ crystal with 46 ppm Fe [5.82]

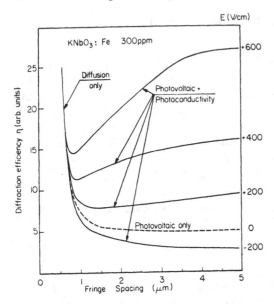

Fig. 5.24. Fringe spacing dependence of the diffraction efficiency of laser-induced gratings in KNbO$_3$:Fe for different applied electric fields [5.79I

bution increases with increasing externally applied electric field, whereas the photovoltaic current does not depend on these quantities. This is shown in Fig. 5.24 for photorefractive gratings formed in KNbO$_3$:Fe 300 ppm [5.79] where the steady-state diffraction efficiency has been plotted.

It has been shown in Sect. 3.6 that the photo-induced gratings in electro-optic crystals may be phase-shifted with respect to the intensity grating. Such stationary shifted gratings can produce an intensity redistribution between interfering beams. The direction of the intensity transfer depends on the sign of the charge-carriers responsible for the space-charge built-up and on the sign of the electro-optic coefficient. A measurement of beam amplification in two-wave mixing experiments therefore allows to determine the sign of carriers if the sign of electro-optic coefficients is known, or vice versa. If charge carriers of different polarity – electrons and holes – contribute, e.g., to diffusion, the current density is reduced compared to single-species diffusion. This results from migration of electrons and holes in the same direction along the concentration gradient. Therefore the value of the photo-induced space-charge field and hence the refractive index change are also reduced, which allows the determination of the ratio of electrons to holes involved in the transport process if the sign of the electro-optic coefficient is known. The dominant polarity can be obtained from the phase shift between the refractive index and the light patterns, see (3.89). In this way, the nature of charge carriers in LiNbO$_3$:Fe and KNbO$_3$:Fe has been determined by optical means [5.80, 81]. In LiNbO$_3$, it has been shown by this technique that both electrons and holes contribute to the grating formation and that the electron contribution dominates in crystals with high Fe^{2+}/Fe^{3+} ratios, whereas hole conductivity predominates in crystals with low Fe^{2+}/Fe^{3+} ratios (Fig. 5.25).

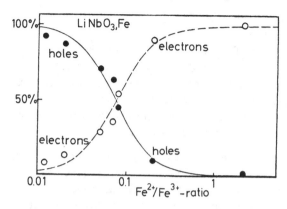

Fig. 5.25. Electron and hole contributions to photoconductivity versus Fe^{2+}/Fe^{3+} ratio in $LiNbO_3$:Fe. (——) and (– – –) represent results of direct photoconductivity measurements, (●) and (○) represent results of the holographic method [5.80]

In photorefractive materials, depending on the crystal symmetry, several components of the refractive indices can be changed by the photo-induced grating. By using the appropriate crystal orientation and polarization of the read-out beam, one is able to determine the relative magnitudes and signs of the electro-optic tensor components. A probe beam that does not affect the space-charge field and is polarized along two optical principal axes of the index ellipsoid can then be used to measure the relative sizes of the electro-optic coefficients describing the refractive index change along these directions. In addition, beam coupling allows one to determine the relative signs of these coefficients.

5.8 Investigations of Photochemical Reactions

By definition, photochemical reactions are induced by light. It is possible to use light also for detection of the reaction because the optical properties of the system usually change during the reaction. A conventional method to detect such changes is absorption spectroscopy, but it has been demonstrated that the transient grating technique offers several advantages for photochemical investigations [5.83–88].

In the grating or holographic technique, two interfering laser beams induce a chemical reaction with a spatial modulation. The resulting optical grating is probed with a third laser beam and the time dependence of the diffraction efficiency gives information on photochemical rate constants, quantum yields and reaction mechanisms.

The advantages of the grating technique compared to absorption spectroscopy are

1) background-free detection of the changes of the optical properties, in contrast to direct absorption measurements where a small change in a large background signal must be detected,

2) high sensitivity due to the background-free detection, and

3) sensitivity to changes in both the index of refraction and absorption. This allows more freedom in the choice of the detection wavelength because the absorption decreases with $(\omega_0 - \omega)^{-2}$ if the probing frequency ω deviates from a characteristic transition ω_0 of a molecule, whereas the refractive index decreases only with $(\omega_0 - \omega)^{-1}$ (Sect. 3.4). The grating technique therefore gives more information in investigations with single-wavelength lasers, whereas tunable sources are usually employed for absorption measurements.

These advantages are, of course, inherent to the grating technique, in general, and not restricted to the investigation of chemical reactions. A drawback of the grating technique is the slightly higher experimental complexity. One might expect also stringent stability and vibration isolation requirements as in high-resolution holography but, for photochemical experiments, one only needs stability comparable to that necessary for any moderate-resolution spectroscopy equipment.

The experiments have been performed with pulsed or stepwise temporal modulation of the laser beams inducing the reaction. Stepwise excitation is realized by simple switching on the excitation laser and observing the temporal growth of the diffracted intensity. The time resolution of experiments performed with this technique was rather poor so far, due to slow laser switching and detection apparatus. There are, however, a number of slow reactions in solid-state photochemistry which take place in the time scale of seconds or minutes and therefore are conveniently investigated. Experiments in fluid solution are also possible. However, in this case, the photochemical grating decays also by mass diffusion. It is still possible to study chemical reactions provided that the characteristic relaxation times are shorter than the diffusion time [5.83], which can be as large as 10 s.

5.8.1 Experiments with cw Lasers (Step-Like Excitation)

A typical experimental arrangement consists of an argon laser to induce the reaction resulting in the grating, a He-Ne laser for probing and a lock-in detection scheme. The description of the experiments given here follows closely the work of *Burland, Bjorklund, Alvarez* and *Bräuchle* [5.84–88].

Simple Photochemical Reactions. For simple photochemical reaction

$$A \xrightarrow{mh\nu} B \, , \tag{5.12}$$

the refractive index change Δn is given by the concentration $C_B = [B]$ of the molecule B

$$\Delta n \propto [B] \, (\delta n_B - \delta n_A) \tag{5.13}$$

where δn_A and δn_B give the refractive index per mole for molecules A and B.

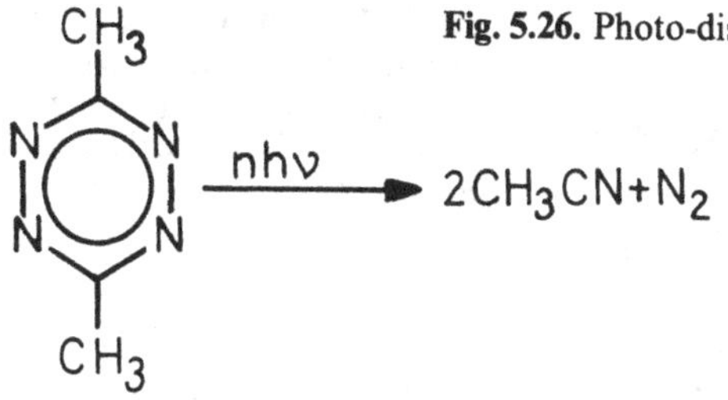

Fig. 5.26. Photo-dissociation of dimethyl-s-tetrazine

Considering early times in the reaction ($[A] \gg [B]$) one obtains

$$[B] = [A]_{t=0} \zeta K I^m t \; , \tag{5.14}$$

where $[A]_{t=0}$ is the concentration of A at time $t=0$, ζ is the quantum yield for the reaction, K the absorption coefficient at the laser wavelength producing the reaction, I the laser intensity and m is the number of necessary photons, e.g., $m=1$ for a one-photon process.

Combination of (5.13, 14) gives the diffracted intensity

$$I_1 \propto (\Delta n)^2 \propto [A]_{t=0}^2 (\delta n_B - \delta n_A)^2 \phi^2 K^2 I^{2m} t^2 \; . \tag{5.15}$$

The diffracted intensity increases quadratically with time t after the laser inducing the reaction is switched on. The number m of necessary photons and the quantum yield can be determined [5.88] from the intensity dependence of the growth curve.

As an example, consider the photo-dissociation of the molecule dimethyl-s-tetrazine (DMST) (Fig. 5.26) in a polyvinylcarbazole (PVK) host. The intensity dependence of the growth curves indicates a two-photon process [5.87]. Similar experiments have been performed with camphorquinone (CQ) in PVK and $m=1$ was obtained. The mechanism of the reaction of CQ in a polymer host is not known in detail [5.85]. For o-nitrobenzaldehyde (ONB) dissolved in polymethyl-metacrylate (PMMA), the number of necessary photons was determined [5.88] to $m=1$ and the photochemical quantum yield to $\zeta=0.17$.

Experiments with Two Excitation Wavelengths. To identify the excited states of a molecule from which reactions take place, the wavelength or energy dependence of the quantum yield can be studied. In the case of multiphoton excitation ($m \geq 2$), two light sources with independent wavelengths are useful. To illustrate the use of the grating technique in this case, the photo-dissociation of carbazole is considered, as shown in Fig. 5.27 [5.87]. Light with a frequency of $\nu_1 \approx 29700 \, \text{cm}^{-1}$ illuminates the sample preparing carbazole molecules in their lowest triplet states T_1. The light with ν_1 does not produce a grating, i.e. a single laser beam or an incoherent lamp is used. If the sample is simultaneously illuminated with two coherent beams at a lower frequency ν_2, a grating can be

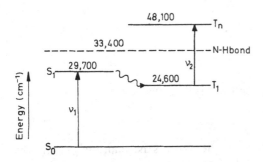

Fig. 5.27. Photo-dissociation scheme and relevant energy levels of carbazole. (---) represents N−H bond dissociation energy

carbazole $\xrightarrow{h\nu_1, h\nu_2}$ carbazyl radical   +H·

produced as a result of the photochemistry that occurs from T_n. It is thus possible to study the photochemistry starting from the higher triplet states without interference by reactions from the lowest triplet state.

Similar experiments have been performed with biacetyl in polycyanoacrylate [5.88]. Here, triplet-triplet absorption extents into the near infrared (600–1100 nm). In combination with UV radiation provided by a Hg-lamp, this material can be used for holographic recording in this spectral region. Chemical development is not necessary, the hologram is fixed, if the UV radiation is switched off and then is not affected anymore by the readout beam.

Current photochemical recording materials in the infrared are not yet sensitive enough to be widely practical. There is therefore great interest to understand the recording process and to identify more sensitive materials. This interest has stimulated much of the work on photochemical reactions described in this section.

Complex Reaction Sequences. Photochemical reactions often occur in several distinct steps where the initial reactants A are converted by absorption of m_1 photons into primary products B. Further absorption of radiation produces secondary products C by means of

$$A \xrightarrow{m_1 h\nu} B \xrightarrow{m_2 h\nu} C \ .$$

An example of such a system is the photochemistry undergone by benzophenone dissolved in PMMA. The probable photochemical reaction is shown in Fig. 5.28. Such a reaction sequence can result in a characteristic non-monotoneous growth-curve of the diffracted intensity I_1 given theoretically by straight-forward extension of (5.13):

$$I_1 \propto (\Delta n)^2 \propto [[B] \delta n_B + [C] \delta n_C - ([B] + [C]) \delta n_A]^2 \ . \tag{5.16}$$

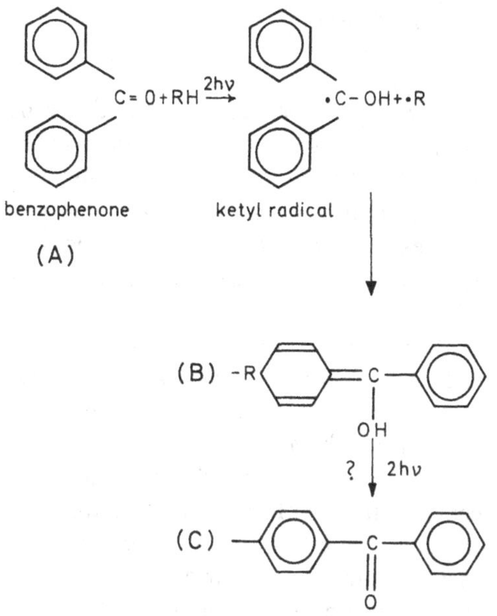

Fig. 5.28. Photochemical reaction scheme of benzophenone

It is assumed now that $\delta n_B > \delta n_A$ and $\delta n_C \ll \delta n_A$. At the initial stages of the reaction: $[C] \approx 0$ and Δn is positive. However, as the reaction proceeds, $[C]$ grows and Δn eventually becomes negative. In between Δn passes through zero and the diffracted intensity vanishes. Such a behaviour has been observed experimentally [5.85] indicating that the total photochemical reaction of benzophenone is at least a two-step process with an intermediate species B.

5.8.2 Pulsed Excitation

The transient grating technique in combination with pulsed excitation has been applied to a number of problems relevant to photochemistry. Relaxation of excited states including rotational diffusion (Sect. 3.4.4), energy transfer (Sect. 5.6), mass diffusion (Sect. 5.2.2) and thermal diffusion (Sects. 3.7 and 5.2.1) are treated in the sections given. Further examples are provided by grating studies of energy deposition in radiationless processes [5.89] and investigations of the cis-trans photochemical isomerization of an azo dye, methyl red [5.83].

In summary, the grating technique has proven to be a simple but sensitive technique for investigating photochemical reactions in solid state materials. In view of the outlined advantages of this method, further applications can be expected. The method has high time resolution. Exploiting this potential, one might hope to follow photochemistry also in liquids.

5.9 Optical Damage and Surface Gratings

When an absorbing target is exposed to two laser beams from different directions, the material is heated with a spatial modulation, as described in Sect. 3.7. If the light is sufficiently intense, this heating can cause melting and evaporation of the material. Permanent gratings have been produced in this way by various authors, e.g. [5.90] and references given in [5.91]. As this monograph primarily deals with dynamic gratings, we stress here transient effects during grating production.

The temporal development of thin-film gratings optically etched with a Q-switched ruby laser has been observed by diffraction of a cw argon laser on a nanosecond time scale [5.91]. Partially transparent aluminium silver and gold films with 50–400 Å thickness on glass substrates have been exposed to energy densities up to 0.5 J/cm^2 in 10–20 ns pulses. The temporal build-up and decay of the diffracted argon laser power has been investigated. The experiments indicate that a spatially periodic plasma of metal ions and slowly moving electrons develops. The relationship between surface damage and surface plasmas has often been discussed and it seems that the grating technique thus offers the possibility to study details of the damage mechanisms.

Similarly the transient grating technique has been applied to study laser annealing of amorphous GaAs [5.92]. Here melting and free-carrier generation lead to subsequent crystallization of the thin film, see Sect. 5.2.2.

For technical production of permanent diffraction gratings, it is advantageous to use cw, visible or near-uv, gas discharge lasers in combination with etching solutions (e.g., HF:H$_2$O, H$_2$SO$_4$:H$_2$O$_2$:H$_2$O or KOH:H$_2$O). The laser-beam illumination of the sample (e.g., Si, GaAs, CdS, InP) leads to a well-resolved local enhancement of the etching rate [5.93, 94]. The time dependence of the grating build-up process has been investigated in [5.95].

Surface gratings or ripples appear also on materials which are illuminated by single laser beams with sufficient power [5.96–103]. These periodic surface structures develop from Fourier components of a random surface disturbance which scatter light from the incident beam very nearly along the surface. The interference of this diffracted optical wave with the incident beam then gives rise to optical interference fringes which can reinforce the initial disturbance [5.100]. The situation is similar to stimulated scattering processes (Sect. 1.2.2) and has been therefore called stimulated surface-polariton wave scattering [5.102]. The dynamics of the grating build-up due to this process have been investigated by observing the dependent diffraction of a probe beam [5.102, 103]. The periodic surface structure can lead to a transition from the initially small (Fresnel) absorption to high absorption and it may therefore be of considerable interest to laser processing of metals and other materials [5.103].

6. Real-Time Holography and Phase Conjugation

In this chapter applications of laser-induced dynamic gratings in real-time holography and phase conjugation are reviewed. The chapter begins with a short introduction to holography and optical computing. Applications of dynamic hologram recording include erasable holographic storage of information, real-time interferometry, real-time recording of optical components such as holographic lenses, prisms etc. and applications in optical signal processing. Nonlinear optical phase conjugation by four-wave mixing in dynamic recording media is discussed. Phase conjugation with cw lasers using photorefractive materials ($BaTiO_3$, $KNbO_3$, ...) and the possibility of self-pumping a phase-conjugate mirror using such materials is described, together with the applications of these mirrors in optical resonators.

6.1 Introduction to Holography

In holography, both the amplitude and phase information of an object wave is stored by using a reference wave and by recording the interference pattern produced by the two beams in the volume of a suitable recording material. The laser-induced dynamic gratings discussed in Chap. 3 are well suited for the storage and processing of the phase and amplitude information of an object with time-varying intensity and position by holographic techniques. In this section, we discuss the basic idea of holography in-so-far as it is required for the understanding of real-time holography experiments with dynamic grating materials. For a more detailed description of holography, we refer the reader to available monographs devoted to this topic [6.1–7].

The basic scheme of holography is shown in Fig. 6.1. In a first step, the interference pattern of the object wave o represented by the electric field distribution:

$$E_o(r, t) = E_o(r) \cos(\omega t - \varphi_o(r)) = \text{Re}\left\{A_o(r)\, e^{i\omega t}\right\} \tag{6.1}$$

and the reference wave R, coherent with o

$$E_R(r, t) = E_R(r) \cos(\omega t - \varphi_R(r)) = \text{Re}\left\{A_R(r)\, e^{i\omega t}\right\} \tag{6.2}$$

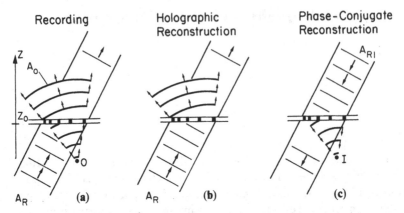

Fig. 6.1a–c. Hologram recording and read-out (principal set-up) demonstrated for recording the spherical wavefront of an object point source; (**a**) recording, (**b**) "normal" read-out by the reference wave A_R produces a virtual image of the object, and (**c**) "phase-conjugate" read-out by illuminating the hologram with a second reference wave A_{RI} antiparallel to the original reference wave A_R produces a phase conjugated object wave and a real image of the object point

is formed in the recording medium, giving the total field

$$E(r, t) = E_o(r, t) + E_R(r, t) \ . \tag{6.3}$$

$E_o(r)$ and $E_R(r)$ are real amplitudes, $\varphi_o(r)$ and $\varphi_R(r)$ are the phases of the object and reference waves, $A_o(r)$ and $A_R(r)$ are complex amplitudes. The power area density within the recording material is then given by

$$I(R, t) = E(R, t)^2/Z = [E_o(R, t) + E_R(R, t)]^2/Z \ . \tag{6.4}$$

R denotes the position vector in the recording material. The optical properties (refractive indices or absorption constant, or both) are assumed to vary with the intensity $I(R)$

$$I(R) = \overline{I(R, t)} = \frac{1}{2Z} |A_o(R) + A_R(R)|^2 = \frac{1}{2Z} (A_o + A_R)(A_o^* + A_R^*) \tag{6.5}$$

in a first approximation, where the bar denotes the time average. Depending on the type of material response, phase or amplitude or tensor gratings are recorded (Sect. 1.1.3). If the recording material is sufficiently thin and the change of the optical properties small, the (amplitude) transmittance of the (recorded) hologram is given by

$$T(R) = \frac{A_t(R)}{A_R(R)} = 1 - \beta I(R) \ , \tag{6.6}$$

where $A_t(R)$ and $A_R(R)$ are the complex amplitudes of the transmitted and illuminating waves, respectively, and β is a real quantity for an amplitude grating (amplitude hologram). For phase grating recording, the phase shift of the transmitted wave φ_t is proportional to the intensity

$$\varphi_t(\boldsymbol{R}) = \gamma I(\boldsymbol{R}) \ , \tag{6.7}$$

which corresponds to a complex transmittance

$$T_p(\boldsymbol{R}) = e^{i\varphi_t(\boldsymbol{R})} = e^{i\gamma I(\boldsymbol{R})} \simeq 1 + i\gamma I(\boldsymbol{R}) \quad \text{for} \quad \gamma I \ll 1 \ . \tag{6.8}$$

Equations (6.6 and 8) show that a combined amplitude and phase hologram is described by (6.6) with a complex β.

In classical holography (e.g., by using photographic plates as the recording medium) the recorded image can be read-out by illuminating the hologram by the reference wave. This second reconstruction step can be performed in two ways: (1) by illuminating the hologram with the same reference beam (Fig. 6.1b, normal reconstruction), or by reading out the hologram with a read-out beam which is antiparallel to the reference beam used during recording (Fig. 6.1c, phase-conjugate reconstruction). For normal reconstruction, the field distribution of the transmitted wave is

$$A_t(\boldsymbol{R}) = T(\boldsymbol{R}) A_R(\boldsymbol{R}) = [1 - \beta I(\boldsymbol{R})] A_R(\boldsymbol{R})$$

$$\simeq A_R(\boldsymbol{R}) - \beta |A_R(\boldsymbol{R})|^2 A_o(\boldsymbol{R}) - \beta A_R^2(\boldsymbol{R}) A_o^*(\boldsymbol{R}) \tag{6.9}$$

$$\text{if} \quad A_o \ll A_R \quad \text{and} \quad \beta |A_R|^2 \ll 1 \ .$$

Equation (6.9) gives the field distribution just behind the hologram. As is known from diffraction theory, the field distribution at a boundary surface determines uniquely the total spatial field. Therefore the first term in (6.9) gives the transmitted reference wave, whereas the second one reconstructs the original reconstructed object wave with an attenuation proportional $\beta |A_R(\boldsymbol{R})|^2$. The third term in (6.9) is the conjugate wave which gives a real ("twin") image of the object if the hologram is thin.

For phase-conjugate reconstruction (Fig. 6.1c) the transmitted read-out wave is given by

$$A_{t|}(\boldsymbol{R}) = T(\boldsymbol{R}) \cdot A_{R|}(\boldsymbol{R}) = T(\boldsymbol{R}) \cdot A_R^*(\boldsymbol{R}) \ , \tag{6.10}$$

$$\simeq A_R^*(\boldsymbol{R}) - \beta [A_R^*(\boldsymbol{R})]^2 A_o(\boldsymbol{R}) - \beta |A_R(\boldsymbol{R})|^2 A_o^*(\boldsymbol{R}) \ . \tag{6.11}$$

The first term again represents the transmitted read-out beam. The second one gives a virtual twin image. Because of the phase factor $\varphi_o(\boldsymbol{R}) - 2\varphi_R(\boldsymbol{R})$, the second term is phase mismatched in a thick hologram, i.e., it will not radiate. The last term in (6.11) represents a wave with conjugate complex amplitude with respect to the object wave. This corresponds to the time-reversed (or phase-conjugate) replica of the original object wave (Fig. 6.1c; see also Sect. 6.9).

In contrast to classical holographic recording materials (silver halide photographic plates, dichromated gelatines etc.) which require three consecutive steps (recording, development and read-out), all these steps occur simultaneously in real-time recording materials, so that any real-time changes of the object wave are continuously recorded.

Hologram recording and readout is possible not only with thin recording materials as outlined above, but also using volume recording where the hologram corresponds to a thick grating (see Sect. 4.3).

The main three attributes to be considered for choosing a dynamic grating material for real-time holography are:

1) Photo-sensitivity,
2) diffraction efficiency, and
3) recording and erasure time.

The recording time constants for the different grating excitations have been discussed in Chap. 3. They range from picoseconds for population density gratings through minutes for the slowest photorefractive media.

In holography, one is primarily interested in the intensity diffracted by the laser-induced refractive index or absorption change. Diffraction from thin and thick amplitude, phase or tensor gratings have been discussed already in Chap. 1.

The holographic diffraction efficiency for thick *phase gratings* (used most often) is given by [6.8], compare also (4.32),

$$\eta = \exp\left(-Kd/\cos\alpha\right)\sin^2\left(\pi\Delta n\,d/\lambda\cos\alpha\right) , \tag{6.12}$$

where d is the hologram thickness, α the Bragg angle corresponding to the wavelength λ, K is the (uniform) absorption constant and Δn the amplitude of the photoinduced phase grating.

The diffraction efficiency for *amplitude gratings* is given by (4.32):

$$\eta = \exp\left(-Kd/\cos\alpha\right)\cdot\sinh^2\left(\Delta K\cdot d/4\cos\alpha\right) , \tag{6.13}$$

where ΔK is the amplitude of the absorption grating. In contrast to phase holograms where the diffraction efficiency can reach 100%, absorption holograms can reach a maximum diffraction efficiency of only 3.7%.

Another important figure of merit for using laser-induced gratings for holographic application is the photosensitivity of a recording material. The photosensitivity may be defined as the incident optical energy required for a given diffraction efficiency and hologram thickness. Alternate definitions give the photoinduced changes of refractive indices or absorption constants for a defined incident or absorbed energy density. Since these definitions are interrelated through (6.12 or 13), we give only the data based on the first definition for a comparison of different recording materials. The photosensitivity is then defined as the change of diffraction efficiency per incident

Table 6.1

Recording Material	Recording or decay time [s]	Required energy density W_0 [mJ/cm²]		Mechanism
LiNbO$_3$	10–10³	3000	[6.9]	
BaTiO$_3$	10–10²	100	[6.9]	photorefractive
KNbO$_3$: Fe²⁺	10⁻⁹–10	10	[6.9]	effect
Bi$_{12}$SiO$_{20}$	10⁻³–1	3	[6.9]	
Dichromated gelatine plates	to be processed	0.01		
High resolution silverhalide photographic plates		10⁻⁵ ($d=10\,\mu$m)		
Rh 6 G-solution	10⁻⁸	30 ($d=100\,\mu$m)ᵃ		excited state population
Si	10⁻⁷	1 ($d=500\,\mu$m)ᵇ		free-carrier excitation
Thermal gratings	10⁻³–10⁻⁶	10ᶜ		temperature change

ᵃ   estimated from Fig. 3.9
ᵇ   calculated from $\eta^{1/2} = \pi \Delta n d/\lambda$ and $\Delta n = n_{eh} K W_0/h\nu$ with n_{eh} and K from Table 3.1
ᶜ   calculated from $\Delta n = (dn/dT)\,(K W_0/\varrho c)$ with $K = (1/d)$ and data for methanol from [6.10]

energy density W_0 for a unit crystal length at the initial stage of hologram formation

$$S = \frac{d(\eta^{1/2})}{dW_0} \cdot \frac{1}{d} \ . \tag{6.14}$$

For a comparison of recording materials, we have listed in Table 6.1 the incident energy density $W_0 = It$ required for $\eta = 1\,\%$ and $d = 1$ mm:

$$S^{-1} \propto W_0(\eta = 0.01, d = 1\,\text{mm}) \ . \tag{6.15}$$

6.2 Real-Time Holography and Interferometry

Laser-induced gratings in dynamic media are ideally suited for real-time holography and interferometry. Materials with high photosensitivity and a large dynamic range (amplitude of the refractive index or absorption grating) or diffraction efficiency are needed for this purpose. Compared to photographic plates, the advantages of dynamic materials for holography are that no chemical processing of the recording medium is needed, and that dynamic holographic gratings follow the changes of the fringe pattern to be recorded. In photorefractive materials, the gratings can be erased by homogeneous illumination. Ultimate recording sensitivities comparable to the ones obtained in high-resolution photographic plates and large diffraction efficiencies can be obtained,

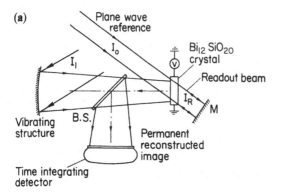

(a)

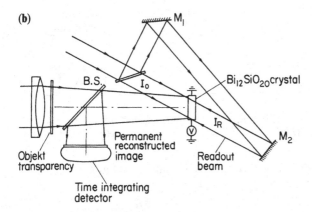

(b)

Fig. 6.2a,b. Optical
configurations for
permanent holographic
recording and read-out for
application to time
averaged interferometry of
vibrating structures:
(a) retroreflected reference
beam, reflecting object,
(b) auxiliary readout beam,
transmission object

e.g., in photorefractive $Bi_{12}SiO_{20}$ or $KNbO_3:Fe^{2+}$ crystals. Laser-induced gratings in these materials are therefore very interesting for real-time holography, double-exposure or time averaged interferometry [6.11].

Optical configurations allowing real-time observation of vibrating structures [6.12] using the time-average interferometry technique are shown in Fig. 6.2. Experiments in double-exposure interferometry can be performed similarly [6.13]. In the optical configuration shown in Fig. 6.2a, the transmitted part of the plane-wave reference beam is used for real-image reconstruction after reflection by mirror M. The beam splitter BS separates the reconstructed beam from the incident illumination. The permanent read-out is destructive, since no wavelength change is allowed for image reconstruction from a volume hologram.

A similar optical configuration is given in Fig. 6.2b. In this case 50 % of the incident reference beam is reflected backwards to the crystal by mirrors M_1 and M_2 for permanent read-out. It has been noted in [6.12] that the configuration of Fig. 6.2a is slightly less efficient than that of Fig. 6.2b, because of the double pass of the reference wave in the crystal and the corresponding increase in the reflection and absorption losses. Nevertheless, the first configuration which cancels the optical activity of $Bi_{12}SiO_{20}$ and $Bi_{12}GeO_{20}$ is simpler for practical use and requires less optical adjustment. Several vibrating objects have been investigated in [6.12–14]. Illuminating these objects in reflection, one obtains a reconstructed image I_d with superimposed interference fringes given by the zero-order Bessel function J_0 [6.1]:

$$I_d \propto \left| J_0 \left(\frac{4\pi}{\lambda} \delta \right) \right|^2 , \tag{6.16}$$

where δ is the local amplitude of periodic deformation. Equation (6.16) is valid if the recording time T_0 of the holographic space-charge field E_{sc} is long compared with the vibration period, i.e., $T_0 \gg T$ for time-average intensity recording. As a consequence of the high sensitivity of photoconductive BSO crystals, T_0, see (3.99), is of the order of 10 ms with 15 mW cm^{-2} incident power. This is convenient for real-time observation of vibrations at frequencies higher than 1 kHz. Lower frequency ranges can be explored by reducing the incident laser power. The optimum parameters for recording holograms of three-dimensional diffuse objects using $Bi_{12}SiO_{20}$ as recording medium have been discussed in [6.14]. Such objects are common, for example, in the field of industrial nondestructive testing of mechanical structures. In a setup similar to the one illustrated in Fig. 6.2a, the authors obtain an optimum beam ratio $I_0/I_1 = 1$ and an optimum time constant of $T_0 = 500$ ms for the hologram build-up of objects having reflectivities between 10^{-3} and 10^{-8}. Therefore the mode pattern of a vibrating structure can be visualized almost instantaneously. This is illustrated in Fig. 6.3 for a vibrating loudspeaker membrane excited at different frequencies. One can clearly see the nodes of the different vibrational modes (bright lines in Fig. 6.3), demonstrating the usefulness of real-time holography for optimum design of vibrating objects. A considerable signal to noise ratio improvement in time-averaged interferometry with $Bi_{12}SiO_{20}$ has been achieved by speckle [6.16] noise suppression with a diffusing screen placed in the object beam and moved step by step between successive exposures on the camera [6.17].

Other special techniques used in holography, such as contour generation of an object (edge enhancement) or speckle interferometry, also work with dynamic grating recording, as is demonstrated in [6.18, 19 and 20].

The edge enhancement is based on the fact that a stored image is a faithful replica of the original object only if the intensity of the object beam is less than the intensity of the reference beam. If this condition is violated, edge enhancement can be produced. The phenomenon of edge enhancement can be understood by considering the modulation ratio or index $m(x)$ of the interfering object $I_1(x)$ and reference beam I_2. If $I_1(x)$ varies slowly compared with the grating spacing Λ, then m can be taken as a local modulation index $m(x)$,

$$(x) = 2 (I_1(x) I_2)^{1/2} / (I_1(x) + I_2) . \tag{6.17}$$

If the object is more intense than the reference beam, then $I_1(x) \gg I_2$, for bright regions of the object, and $m(x)$ will be much less than its maximum value of unity

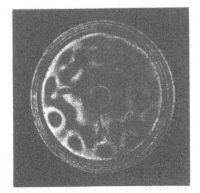

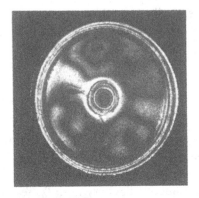

F:2KHz
V_{cc}:0.5v

F:6KHz
V_{cc}:0.5v

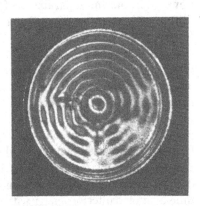

F:10KHz
V_{cc}:5v

F:12 KHz
V_{cc}:7v

Fig. 6.3. Time-averaged interferogram of a diffuse vibrating object: mode pattern visualization of a loudspeaker membrane excited at different frequencies F and voltages V_{cc} [6.12]

(Fig. 6.4). Similarly, $I_1(x) \ll I_2$, for dark regions of the object, and $m(x)$ will still be small (Fig. 6.4). However, $I_1(x_0) = I_2$, at some point x_0 in the transition region between a bright and a dark region in the object (i.e., at an edge), which gives $m(x_0)$ equal to unity; consequently, a local grating with a large diffraction efficiency is produced in the region around x_0. The reading beam will be preferentially scattered by this grating and the corresponding image will therefore have that edge enhanced. Such experiments in dynamic recording materials have been performed successfully with $Bi_{12}SiO_{20}$ [6.20] and $BaTiO_3$ [6.19] crystals. Fig. 6.5 shows examples of edge enhancement using $BaTiO_3$ (a) and $Bi_{12}SiO_{20}$ crystals (b).

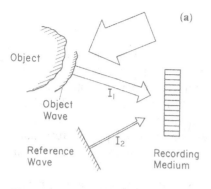

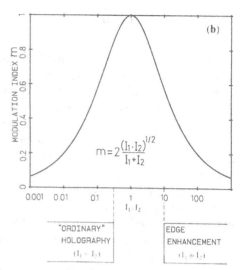

Fig. 6.4. (a) Recording configuration for edge enhancement and **(b)** intensity ratio dependence I_1/I_2 of the fringe modulation ratio m. Edge or contrast enhancement occurs for recording with weak reference beams $I_2 \ll I_1$

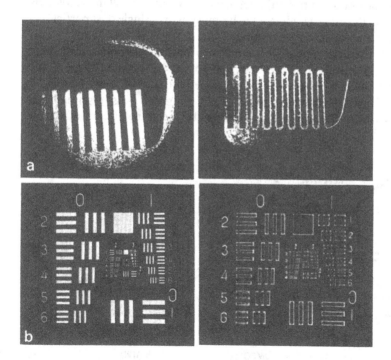

Fig. 6.5. (a) Edge enhancement of a comb produced by nonlinear recording in a photorefractive BaTiO$_3$ crystal [6.19] and **(b)** of a test pattern in a photorefractive Bi$_{12}$SiO$_{20}$ crystal [6.20]

6.3 Information Storage

Optical information storage in the volume of a dynamic medium has the advantage of offering a high capacity of up to 1 bit per cube of recording wavelength (typically up to 10^{12} bits/cm^3), contact-free parallel read-out of an array of information units and the possibility of erasing part or the whole memory for writing-in new informations. Besides the bit-by-bit recording presently used also for recording data in video or audio discs, parallel recording of complete pages of information in the form of holograms has been demonstrated.

Hologram storage of an array of bits offers many advantages as compared with the bit-by-bit storage method, where single bits are recorded at different places of the recording medium. Thus a major advantage of holographic storage is that it requires a much smaller mechanical precision and that it is less sensitive to imperfections of the storage medium. Since the Fourier transform and not the bit pattern itself is stored, the information about each bit is spatially spread over the entire hologram [6.21]. Dust and scratches would decrease to some degree the signal-to-noise ratio for the entire set of bits rather than result in the total loss of some bits. On the other hand, in a bit-by-bit system, a single dust particle can easily wipe out one or more bits entirely.

Let us now consider a possible arrangement for a holographic read-write-erase memory. A schematic of the basic components is shown in Fig. 6.6. It is characteristic for this kind of system that large blocks of data are handled in parallel. The input element – called page composer – is essentially an electronically controlled transparency where the incoming stream of data is

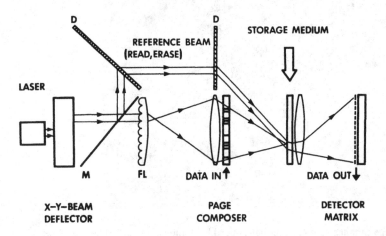

Fig. 6.6. Conceptional model of a holographic memory consisting of laser source, beam deflector, page composer, storage medium and detector matrix. A fly's eye lense (FL) enables guiding of the signal beam to the different subhologram areas. A mirror (M) and diffraction gratings (D) ensure synchroneous movement of the reference beam [6.21]

converted into an array of dark and light areas representing the binary zeros and ones of a page of digital data. Coherent light from a laser illuminates the page composer, and the transmitted light – called the object beam (or signal beam) – interferes with a reference beam at a well defined position in the dynamic grating medium. The resulting interference pattern produces the photo-induced refractive index or absorption changes. The data page is thus stored in the form of a phase or amplitude hologram. When the recording of one page is completed, a different position on the storage medium is addressed with the help of a beam deflection system in conjunction with suitable optics, such as a fly's eye lens. Precise overlap of the signal beam and the reference beam must be maintained when the beams move to a new position. Finally, the entire storage medium will be covered by an array of holograms each representing a data page with a large number of bits.

For read-out of any of the stored data blocks the desired hologram is addressed by the reference beam alone. The light diffracted from the hologram is a faithful replica of the original wavefront from the page composer. A matrix of light detectors – one for each raster position – is placed in the image plane, and the content of the hologram is reconverted into a sequence of electrical signals. The reconstructed images from each of the holograms will appear at the same position – the position of the optical image of the page composer. Therefore the same detector arrangement can be used for read-out of all the different holograms without having to move the detector matrix.

A very attractive feature of a holographic system is the possibility of volume storage. In a photosensitive medium where bulk changes of n are produced, many holograms can be superimposed in the same element of volume [6.21]. This is accomplished, for example, by varying the angle of incidence of the reference beam, each individual hologram being associated with a well defined angle of incidence. Selective reconstruction of the superimposed holograms can be achieved because efficient diffraction takes place only when the hologram is addressed at the Bragg angle. Volume storage is clearly the key to very high bit packing densities and consequently to high capacity and speed.

A single hologram of about 1 mm^2 area may contain 10^4 bits or more [6.22]. Presently up to 500 holograms can be superimposed reversibly at each site [6.23, 24] in LiNbO$_3$ crystals, giving a total capacity of 0.5 Gbits/cm^2. The access time to any of the data blocks depends primarily on the time it takes to switch the light beam from one position to another, e.g., a few microseconds [6.25]. Therefore random access times in the microsecond regime and data read-out rates of the order of 10^9 bits/s appear feasible, provided, of course, that the other system components do not limit the speed.

Due to the rapid advent of high-density semiconductor memories which interface well with conventional electronic technologies, the interest in holographic storage has descreased in recent years. At the present time, there is little doubt that, for completely reversible write-read-erase memories of not too high capacity, nonoptical devices such as semiconductor memories will dominate. However, for very large capacities, difficulties are expected due to the limitation of the number of bits that can be placed on a single chip (the state of the art is about 10^6 bits). To increase the total capacity, a very large number of individual chips must be interconnected, and this problem will eventually dictate a practical upper limit for the total capacity. It appears, therefore, that apart from specialized applications the real impact of optical memories will be in the capacity range of 10^9 bits and more, where the combination of high storage density and high speed will be a decisive factor. However, none of the materials developed up to now completely satisfies all the requirements of memory elements. While dynamic grating recording offers the potential of a read-write-erase memory, the major drawbacks are that the grating storage times of high-sensitive materials are short at room temperature and that erasure of a single bit of information requires the erasure of the entire memory contained in the illuminated volume. Materials that exhibit recording sensitivities comparable with high resolution silver halide emulsions, e.g. $KNbO_3$, $K(NbTa)O_3$ and $Bi_{12}SiO_{20}$, have storage times in the dark at room temperature of only a few hours. Even in doped $LiNbO_3$, storage times are only a few months at best. Only in $LiTaO_3$ storage times of 10 years have been reached. Increasing the storage time by decreasing the ambient temperature does not seem attractive. A further serious complication arises because the Bragg condition for thick holograms requires that the same wavelength of light must be used for read-out and writing. In the absence of any fixing process, the material remains photosensitive to the read-out beam, resulting in degradation of the stored image during read-out. Of course, this degradation is most serious in the most sensitive materials. One approach to overcome this problem has been to fix the image, rendering it nonvolatile (or less volatile) as described below, but now the versatility of the memory is decreased since the image can no longer be optically erased.

6.4 Holographic Lenses and Gratings

Real-time optical components such as lenses, prisms or mirrors can be realized in dynamic media by holographic wavefront conversion. A prism will deflect a plane wave into another plane wave propagating in a different direction and can be realized with a simple grating recorded by two plane waves. Cylindrical lenses may be produced by recording a hologram with a plane and a cylindrical wave, or with two cylindrical waves. For three-dimensional lenses and mirrors, the cylindrical waves need to be replaced by spherical waves [6.15, 26].

Figure 6.7a shows the production of a holographic lens-prism combination. Figure 6.7b demonstrates the use of this holographic element to produce a collimated beam from a glass fiber source. Two holographic lens-prism combinations can be used to couple two optical fibers, as indicated in Fig. 6.7f. Such a wavefront conversion scheme is useful for coupling fibers with cylindrical or elliptical cross section or thin film optical waveguides as discussed in [6.2, 27–31].

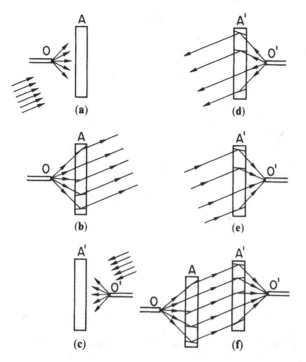

Fig. 6.7. (a) Recording of a hologram with a plane wave and radiation from O; (b) reconstruction of the plane wave by radiation from O; (c) recording of a hologram with a plane wave and radiation from O'; (d) reconstruction of the plane wave by radiation from O'; (e) the conjugate of the plane wave in (d) converges upon O'; (f) a holographic coupler between the fibres at O and O' [6.2]

As compared to conventional optics, holographic components may have the advantages that they might perform several functions in a single element, and that the same function may be performed easily in parallel, resulting in an array of lenses (fly's eye array).

6.5 Wavelength Conversion of Optical Images

Several methods for optical frequency conversion of three-dimensional images using real-time holography have been investigated [6.32, 33]. The wavelength conversion is obtained by reading-out the dynamic holograms at a wavelength different from the one used for recording λ_1. Infrared to visible up-conversion with a demagnification of the (infrared) image at λ_1 is obtained by using visible read-out light (λ_2). Magnification of $M = \lambda_2/\lambda_1$ occurs if holograms are recorded in the UV at λ_1 and read-out at λ_2 (e.g., in the visible).

The main requirement for a high-quality wavelength conversion of two- or three-dimensional objects is that the interacting beams must fulfill certain phase-matching or wavevector conservation conditions (Bragg conditions) for all spatial frequencies involved in the hologram. This can be done in a nonresonant way by employing a phase-matched four-wave mixing process. With such a technique, *Martin* and *Hellwarth* [6.33] have successfully up-

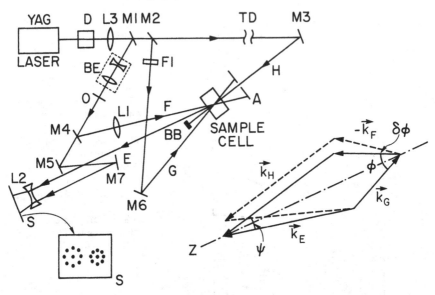

Fig. 6.8. Schematic of apparatus used to observe up-converted phase-conjugated images, and the phase-matching diagram for the waves involved. Infrared beam F, modulated by the object O, beats with the infrared beam G and the visible beam H in a sample to generate a visible phase-conjugate replica-beam E which is observed (alongside reference fraction of F) at S. D is a frequency doubler, BE a beam expander, Fl is a 532 nm blocking filter, BB is a beam block, A is an aperture, S is a screen or film paper, TD is a variable time delay, and O is a hole pattern. (M: mirror, L: lens) [6.33]

converted $\lambda_1 = 1.06\,\mu m$ images into the visible with $\lambda_2 = 0.53\,\mu m$ by using photoinduced population density and thermal gratings in several liquids. The set-up used for these experiments is shown in Fig. 6.8. A hologram of the object O is recorded by interfering the object beam F with a reference beam G both at $\lambda_1 = 1.06\,\mu m$ in the sample cell. This hologram is read-out by the frequency-doubled Nd:YAG laser at $\lambda_2 = \lambda_1/2 = 0.53\,\mu m$ incident to the hologram to fulfill

$$k_E \equiv k_G - k_F + k_H \,, \qquad\qquad (6.18)$$

where k_i are the wavevectors of the different beams involved, and k_E is the outcoming up-converted signal beam. It was shown in [6.33] that the input wavevector k_F and signal beam k_E can rotate by a relatively large angle without spoiling phase matching over a typical beam interaction length (≈ 1 mm).

Therefore an infrared reference image O in the F beam could be frequency shifted by this four-wave mixing experiment, to yield a demagnified up-converted replica of the image (Fig. 6.9). The maximum conversion efficiency reached in the experiments was 5% with an infrared absorbing dye (Kodak No. 1401S Q-switching dye) added to 1–2 dichloroethane.

The number N of image resolution elements that can be converted by the geometry of Fig. 6.8 is approximately the solid angle Ω_p in which k_E is phase matched, divided by the diffraction solid angle Ω_d. Phase matching is obtained

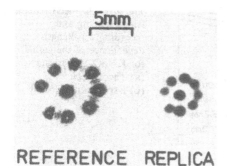

5mm

REFERENCE   REPLICA

Fig. 6.9. Photographic images of the infrared-image (reference) beam F and the phase-conjugated (replica) beam E obtained by placing ordinary photographic print paper at the observation plane S in Fig. 6.8. The reference image is made by the green light from the 4% reflectance of M1 in Fig. 6.8, the print paper being insensitive to infrared. The Fresnel fringes seen on the reference beam pattern should be, and are, absent from the phase-conjugate pattern at the plane of observation [6.33]

when $\Delta k \equiv |k_E - n_r \omega/c| \leq \pi/l$. Here, l is the effective length over which the nonlinear polarization is generated; it is given by the beam overlap geometry, sample length, or the infrared absorption length, whichever is least. An examination of the geometry of Fig. 6.8 shows k_E is unchanged when k_F is rotated about the symmetry (Z) axis of the k-vector diamond. That small deviation $\delta\phi$ of ϕ (about its average value ϕ_0) which makes $\Delta k\,l = \pi$ is easily seen to be such that (for $\phi_0 \ll 1$) $\Omega_p \simeq 2\pi\phi_0\delta\phi \simeq 4\pi^2(k_F l|1 - k_F/k_E|)^{-1}$. Taking the diffraction solid angle of the input beam to be $4\pi^2/k_F^2 A$, A being the beam area, one gets for the number N of phase-matched image resolution elements [6.33]

$$N \simeq \frac{Ak_F}{l|1 - k_F/k_E|} \ . \tag{6.19}$$

As k_F approaches k_E, this number approaches infinity. This means that, for degenerate four-wave mixing in the geometry of Fig. 6.8, phase matching is satisfied for all input beam angles (and the generated E wave is the time-reversed replica of the F wave).

6.6 Image Amplification

Another application of dynamic holography is the amplification of coherent light beams, including amplification of beams carrying optical information. The effect makes use of the beam coupling occurring during the recording of dynamic volume holograms. It has been shown in Sect. 4.5 that two-wave mixing in recording materials which allow for a fringe mismatch between the intensity and the refractive index grating leads to an energy redistribution between the recording beams, thus transferring optical energy from one beam to the other. This energy transfer can be characterized by an exponential gain factor Γ.

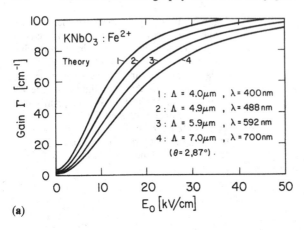

(a)

Fig. 6.10a, b. Electric field, fringe spacing and recording wavelength dependence of the gain Γ for $KNbO_3$: Fe crystals (a) Theoretical, (b) Experimental [6.34]

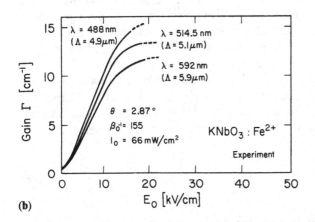

(b)

Extending the theory in Sect. 4.5.3 to include absorption and arbitrary grating phase shift φ, the amplified (signal) intensity is given by

$$I_1 = \frac{I_0 \beta_0}{1 + \beta_0 \exp \Gamma d} \exp (\Gamma - K) d \underset{(\beta_0 \ll 1)}{\simeq} I_{10} \exp (\Gamma - K) d \, , \qquad (6.20)$$

where K is the light absorption constant, $I_0 = I_{10} + I_{20}$ the total intensity incident to the crystal, $\beta_0 = I_{10}/I_{20}$, I_{10} and I_{20} the incident, and I_1 and I_2 the transmitted intensities. For $K = 0$ Eq. (6.20) reduces to (4.62). The gain Γ is given by [6.9],

$$\Gamma(x) = \frac{4 \pi \beta_0(x) \Delta n_1(x)}{\lambda \cos \alpha} \sin \varphi(x) \, , \qquad (6.21)$$

where $\Delta n_1(x)$ is the amplitude of the photoinduced grating and φ the grating phase shift.

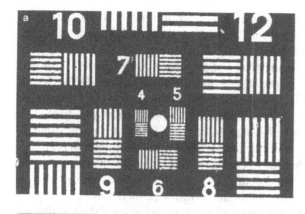

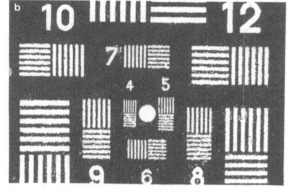

Fig. 6.11. Test object imaged through amplifier crystal without pump beam (**a**). With a pump beam, the test object is amplified ca. 4000 times (**b**) after *Laeri, Tschudi, Albers* [6.35]

The dynamic nature of the beam coupling during the two-wave mixing process is such that it can be used to amplify time-dependent signals. Figure 6.10a shows the theoretical dependence of the gain Γ on the applied electric field for photorefractive recording in $KNbO_3 : Fe^{2+}$. The increase of Γ with increasing field strength is due to an increasing fringe mismatch φ for higher field strengths, which reaches $\varphi = \pi/2$ for $E \simeq 18\,kV/cm$ ($qL_E \simeq 1$; $q = 2\pi/\Lambda$ grating wavevector, $L_E = \mu\tau E_0$: drift length) and saturates. Figure 6.10b shows the experimental dependence of the gain Γ on the electric field strength.

Very large gain factors $g = \exp(\Gamma - K)d$ of up to 4000 have been reached in a 3.2 mm long $BaTiO_3$ crystal and higher values are possible in larger crystals with optimized gain (e.g., for $\Gamma - K = 11.5\,cm^{-1}$ obtainable in $KNbO_3$ as shown in Fig. 6.10b and $d = 1\,cm$, $g \gtrsim 10^5$). The fast recording time of $T_0 \simeq 10\,ms$ for an intensity $I = 155\,mW/cm^2$ and the possibility of controlling the gain Γ in photoconducting materials by an external electric field [6.36] offers further advantages in coherent image processing applications. Optical elements such as optical limiters, pulse shapers, differential amplifiers, optical triodes, optical switches and logical elements proposed in [6.37] using the more complicated arrangement of an electrooptic crystal within a Fabry-Perot resonator could be realized with dynamic grating materials.

The above-mentioned technique of image amplification is not limited to plane images but has already been applied to the amplification of three-dimensional images [6.38]. However, in the experiments reported in [6.38], the recording material $LiNbO_3$ was not very photo-sensitive and therefore special techniques had to be used for recording the diffuse volume object. Materials with higher sensitivities ($Bi_{12}SiO_{20}$, $Bi_{12}GeO_{20}$ and reduced $KNbO_3$) make the recording and amplification of three-dimensional objects more practical.

The possibility of coherent image subtraction and addition using $LiNbO_3$ will be discussed in Sect. 6.7. To perform these and a number of other operations, a classical two-exposure technique of complex amplitude subtraction has to be used [6.39]. Using coherent light amplification, the same operations can also be performed with a single exposure technique, by introducing the second image in the reference beam. The common parts of the two images are then added or subtracted in the same manner as in [6.34]. Besides the advantage of having a single exposure, the beam coupling technique is an intensity subtraction method and insensitive to phase inhomogeneities of the object beam.

6.7 Image Subtraction, Logical Operations

A further application of photoinduced gratings in real-time information processing is image subtraction or addition. Superimposed holograms recorded in dynamic media can be selectively erased. Depending on the recording and erasure configuration, a series of Boolean operations can be performed.

The principle of selectively erasing a complete hologram (or part of it) consists of superimposing a complimentary refractive index spatial modulation

$$\Delta n_1(x, y, z) = \delta n_1 \cos [qx + \phi_1(x, y, z)] \quad \text{on}$$

$$\Delta n(x, y, z) = \delta n \cos [qx + \phi(x, y, z)]$$

(6.22)

of the stored hologram such that $\Delta n_1(x, y, z) = -\Delta n(x, y, z)$. This can be performed by a π-phase shifted reference wave during a second recording of the object to be erased, provided that the photoinduced change Δn is proportional to the incident light intensity. The feasibility of this process has been demonstrated for Fourier holograms recorded in photorefractive $LiNbO_3$ [6.39, 40]. Selective erasure of a page of information (complete, partial, and single-bit erasure) using this technique is reproduced in Fig. 6.12. For partial erasure of a certain information block or bit, a partially masked transparency was used for recording that unit with a π-shifted reference beam.

Superimposing two different binary variables A and B in this way into the same hologram corresponds to the Boolean operation "Exclusive Or" ($\circ$):

$$A \circ B = (A \cap \bar{B}) \cup (\bar{A} \cap B) .$$

(6.23)

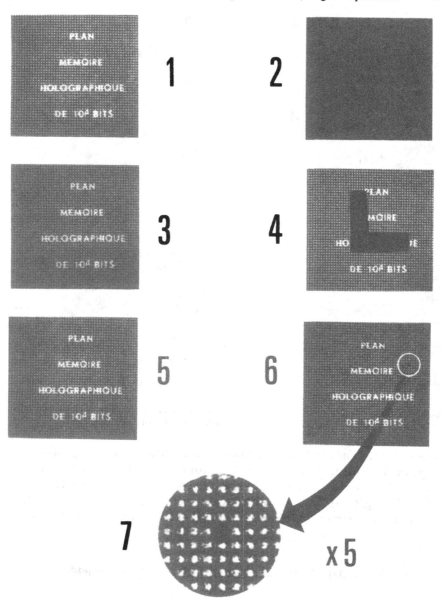

Fig. 6.12. Selective erasure of a complete (1, 2), or partial page (3, 4) or a single bit (5, 6, 7) in photorefractive LiNbO$_3$: Fe [6.41] (reproduced with courtesy of J. P. Huignard, Thomson – CSF)

Logical Operations

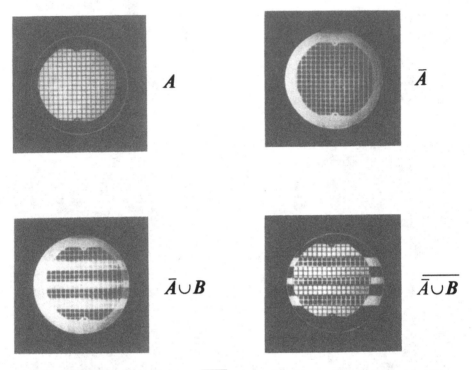

Fig. 6.13. Logical operations ($\bar{A} \cup B$ and $\overline{\bar{A} \cup B}$) between two binary transparencies A and B in photorefractive LiNbO$_3$: Fe [6.41] (reproduced with courtesy of J. P. Huignard)

where $\cap$ and $\cup$ stand for the intersection (and) and disjunction (or) operators. In the particular situation where B is a uniform object transparency ($B=1$), the reconstruction of the resulting image is the object A with reversed contrast ($A \circ 1 = \bar{A}$) (Fig. 6.13). This and other logical operations ($\bar{A}$, $\bar{A} \cup B$, $A \cup \bar{B}$ etc.) have been demonstrated by *Huignard* et al. [6.41] and are shown in Fig. 6.13.

6.8 Spatial Convolution and Correlation of Optical Fields

Dynamic cross correlation or spatial convolution with a classical Fourier transform optical configuration can be achieved by two-or four-wave mixing of two monochromatic wave fields in a dynamic recording material [6.11, 42]. Spatial convolution and correlation of monochromatic fields by four-wave mixing in the common focal plane of a two-lens system in photorefractive materials has been reported in [6.43]. The four-wave mixing geometry used for these experiments is shown in Fig. 6.14. Fields 1 and 4 propagate essentially in the x direction, while 2 propagates in the $-z$ direction. All three fields, which are

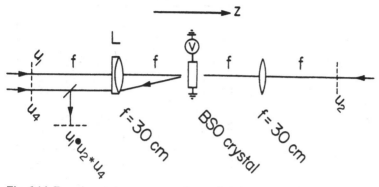

Fig. 6.14. Experimental arrangement for performing convolution and correlation. Input and output planes are shown by (---) [6.43]

of the same frequency, may contain arbitrary spatial modulation which is different for each wave. The input amplitudes $u_1(x,y)$, $u_2(x,y)$, and $u_4(x,y)$ in the outer focal planes are Fourier transformed (FT) by propagating to the common focal plane, The transformed fields $U_1 = FT u_1(x,y)$, $U_2 = FT u_2(x,y)$ and $U_4 = FT u_4(x,y)$ are incident upon the dynamic recording medium of thickness d. The exact Fourier transform only exists in the back focal plane. Therefore, there is an upper limit for the thickness d of the processing element, for which the transformation is still satisfactory. This is the case for

$$d < 2f^2 \cdot \lambda/r_{max}^2 \ , \tag{6.24}$$

where r_{max} is the largest spatial extent of all u_i [6.43].

The convolution theorem [6.9], which states that the Fourier transform of a product equals the convolution of the factors, implies that the backward wave u_3, when evaluated at the plane located a distance f in front of lens L, is of the form

$$u_3(x_0, y_0) = \psi u_1(-x, -y) \bullet u_2(-x, -y) * u_4(-x, -y) \ , \tag{6.25}$$

where ψ depends on the nonlinear susceptibility of the material and on the parameters of the Fourier-transform lens and the symbols $\bullet$ and $*$ denote convolution and cross correlation integrals [6.45].

Two-image processing has been performed in [6.43] by simulating a δ-function for the third input beam u_2 or u_4 (see Fig. 6.15). The simulation is accomplished by placing a third lens outside the two-lens system so that it focuses at the third input plane and gives a collimated beam at the transform plane. For objects with real amplitude transmittance u_i, the correlation and convolution operations differ only in the sign of the arguments of u_i [6.11] so that the correlation can be achieved by convolving the inverted object and vice versa.

Typical experimental results for amplitude objects mixed in $Bi_{12}SiO_{20}$ are shown in Fig. 6.15. The first three columns list the input fields u_1, u_2, and u_4. The

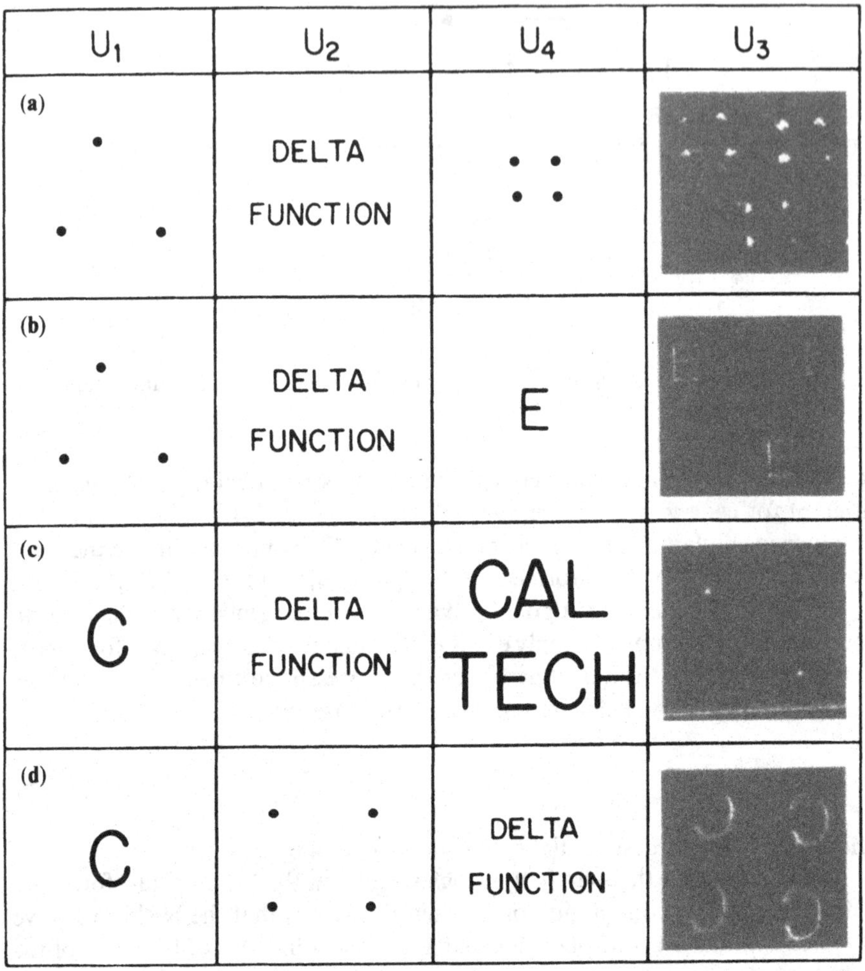

U_1	U_2	U_4	U_3
(a) • • •	DELTA FUNCTION	• • • •	
(b) • • •	DELTA FUNCTION	E	
(c) C	DELTA FUNCTION	CAL TECH	
(d) C	• • • •	DELTA FUNCTION	

Fig. 6.15. Real-time correlation and convolution in photorefractive $Bi_{12}SiO_{20}$. First three columns are the input objects u_1, u_2 and u_4. The last column are photographs of the output u_3 after *White, Yariv* [6.43]

fourth column shows the photographs obtained at the output plane. Rows (a–c) illustrate correlation and row (d) illustrates convolution. It has been demonstrated in [6.43] that not only amplitude objects can be processed but also purely phase objects.

Real-time cross correlation by two-wave mixing of optical beams, which are spatially modulated by photographic transparencies, has been reported using saturable absorbers [6.45] or $Bi_{12}SiO_{20}$ crystals [6.46]. In a classical two-lens coherent optical processor, *Pichon* and *Huignard* have simultaneously recorded and read out a dynamic phase-volume hologram using very low optical power levels ($I_0 \simeq 200\,\mu$W at $\lambda = 488$ nm) and retaining a very low cycle time to build up or erase the correlation peak ($T_0 \simeq 25$ ms). However, further work has to be done in order to study the limitations which arise on the space-bandwidth product with this arrangement in applications such as pattern recognition, optical computing and object tracking.

6.9 Phase Conjugation and Time Reversal

Nonlinear optical phase conjugation by four-wave mixing in dynamic recording media has undergone a tremendous development in the last few years [6.47–54]. Much of the interest in this application of laser-induced gratings arises from the fascination with the idea of time reversal of an electromagnetic wave and the demonstration of restoration of severely distorted optical beams into their original, non-aberrated state [6.53, 54]. In order to understand the properties of conjugate fields, we consider an optical wave of frequency ω propagating in the $+z$ direction

$$E_1(r, t) = |A(r)| \cos\left[\omega t - kz - \varphi(r)\right] = \operatorname{Re}\left\{A(r)\mathrm{e}^{\mathrm{i}(\omega t - kz)}\right\} , \qquad (6.26)$$

where $A(r)$ is a complex amplitude containing the phase information $\varphi(r)$. The time-reversed wave $E_2(r, t) \equiv E_1(r, -t)$ is then given by

$$E_2(r, t) = |A(r)| \cos\left[-\omega t - kz - \varphi(r)\right]$$

$$\equiv |A(r)| \cos\left[\omega t + kz + \varphi(r)\right] = \operatorname{Re}\left\{A^*(r)\mathrm{e}^{\mathrm{i}(\omega t + kz)}\right\} , \qquad (6.27)$$

where $A^*(r)$ is the complex conjugate of the complex amplitude $A(r)$. This conjugate wave corresponds to a wave moving in the $-z$ direction, with the phase $\varphi(r)$ reversed relative to the incident wave. Phase conjugation can thus be regarded as a type of reflection combined with phase reversal. It is equivalent to leaving the spatial part of E unchanged and reversing the sign of t; in this sense, phase conjugation is equivalent to time reversal.

An illustration of phase conjugation can be obtained by comparing the reflections from an ordinary and from a conjugate mirror. As Fig. 6.16 shows, a diverging spherical wave striking an ordinary mirror at an angle Θ leaves it at an angle $-\Theta$ and continues to diverge. In contrast, the same wave is reflected from the conjugate mirror as a converging wave that retraces the incident light path to the point where the wave originated. Therefore light originating from any three-dimensional object will be mapped exactly onto the object. If a semitransparent mirror is placed between the object and the phase conjugate mirror, part of the object can be imaged to another place (lensless imaging).

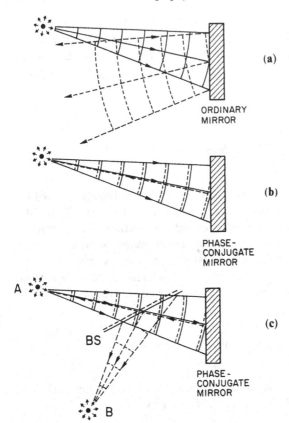

Fig. 6.16. Comparison of the reflection of a spherical wave from an ordinary (**a**) and a phase conjugate (**b**) mirror. (**c**) Optical imaging by a phase-conjugate mirror by means of a beam splitter (BS) which deflects the retroreflected phase conjugate wave of an object A to yield an image B

(a)

ORDINARY
MIRROR

(b)

PHASE-
CONJUGATE
MIRROR

A

BS

(c)

PHASE-
CONJUGATE
MIRROR

B

The propagation laws of electromagnetic fields imply that the phase-conjugate wave retraces back the light path of the incident field. In particular, any wavefront distortion of an incident waveform experienced by an aberrating medium (nonuniform glass, turbulent atmosphere or liquids, or irregular optical fibre, etc.) will be transmitted (with phase reversal) to the reflected beam by the phase-conjugate mirror, and will thus be rectified after the second transit in the aberrating medium (Fig. 6.17). Provided that the recording time of the laser-induced gratings used for the generation of the conjugate wave is sufficiently short and that the phase disturbance is unchanged within the propagation time of the light wave from the disturbance to the phase conjugator and back, any time varying disturbance can be compensated for by the phase conjugator.

To show that the backward propagating conjugate wave is a valid solution of the wave equation, we can use the following argument. For simplicity, we assume a single polarization component of the probe field of the form given by (6.26).

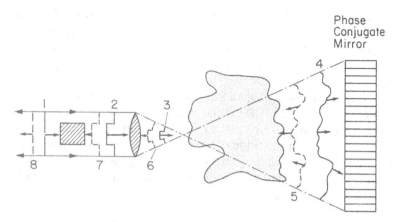

Fig. 6.17. Compensation of phase wavefront alteration from optical elements and phase aberrators by means of optical phase conjugation. A monochromatic plane wave incident from the left side (*full lines*) (1) is distorted (2–4) as it passes a prism (2) a lens (3) and a region with nonuniform refractive index $n(R)$ (4). The resultant distorted wavefront is conjugated (time reversed) by the phase conjugate mirror, and is restored after retraversing the same optical elements as the original plane wavefront (5–8)

The amplitude of the signal wave $E_1(r, t)$ propagating in a medium with refractive index $n(r)$ is then a solution of the scalar wave equation

$$\nabla^2 E_1 - \frac{n^2(r)}{c^2} \frac{\partial^2 E_1}{\partial t^2} = 0 \ , \tag{6.28}$$

where c is the speed of light in vacuum.

Inserting (6.26) into (6.28) and using the slowly varying envelope approximation

$$|k^2 E_1| \gg \left| k \frac{\partial E_1}{\partial z} \right| \gg \left| \frac{\partial^2 E_1}{\partial z^2} \right| \ , \tag{6.29}$$

one gets

$$\nabla^2 A + \left[\frac{\omega^2}{C^2} n^2(r) - k^2 \right] A - 2ik \frac{\partial A}{\partial z} = 0 \ . \tag{6.30}$$

Under the assumption that $n(r)$ is real (no absorption!) we get for the conjugate complex of (6.30):

$$\nabla^2 A^* + \left[\frac{\omega^2}{C} n^2(r) - k^2 \right] A^* + 2ik \frac{\partial A^*}{\partial z} = 0 \ , \tag{6.31}$$

which is recognized to be the wave equation describing the propagation of A^* in the direction $-z$. Hence the conjugate wave $E_2(r, t)$ of $E_1(r, t)$ satisfies the wave equation, and therefore describes the propagation of a field having the same equiphase surfaces as $E_1(r, t)$ at each point in space and propagating in a direction opposite to that of the incident wave $E_1(r, t)$. If we assume that the phase-conjugate mirror is located at a plane $z = z_0$, then the above arguments hold for all $z \leq z_0$.

The application of phase conjugation in adaptive optics, where aberrated wavefronts should be restored to their initial state, is illustrated in Fig. 6.17. An incident plane wave front passing through homogeneous optical elements is changed, passes through an additional inhomogeneous spatially dependent phase aberrating medium, resulting in distorted equiphase surfaces. After interaction with the phase conjugating mirror and subsequent passage back through the same aberrator and optical components, the initial planar equiphase surfaces are recovered.

The generation of the phase-conjugate wave occurs through nonlinear mixing of optical waves in a dynamically recording medium. One possible scheme, which is directly related to the recording scheme used in holography (Fig. 6.1), is four-wave mixing. It has been shown in Sect. 6.1 that a conjugate wave A_o^* is generated by mixing a signal wave A_o to be conjugated with two antiparallel pump waves A_R and $A_{R|}$ (Fig. 6.1) (degenerate four-wave mixing). Therefore any dynamic recording medium with photoinduced changes of refractive indices or absorption constant can be used as phase conjugator in this configuration.

6.10 Applications of Phase Conjugation

In this section we shall discuss two applications of optical phase conjugation: lensless imaging and focussing, and real-time compensation of phase-distorted optical waves which have passed an aberrating medium. Both applications make use of the fact that the phase-conjugate wave exactly retraces the light path of the original wave.

6.10.1 Lensless Imaging and Focussing

The principle of lensless imaging and focussing was already described in Sect. 6.9 and Fig. 6.16. Because the phase-conjugate wave exactly retraces the light path of the original wave, each image point is exactly mapped (distortion-free) to the origin (or some other location if a partially transparent beam splitter is inserted in the light path). Lensless imaging has been demonstrated in photolithography with photorefractive LiNbO$_3$ [6.55] or ruby [6.56] as the phase-conjugator material. In the experiments using a LiNbO$_3$ crystal, the configuration shown in Fig. 6.18 was used. The photorefractive crystal was illuminated by the pump waves and the signal wave at $\lambda = 413$ nm (Kr$^+$ laser). The signal wave originated from an illuminated resolution test-pattern drawn by electron-beam lithography

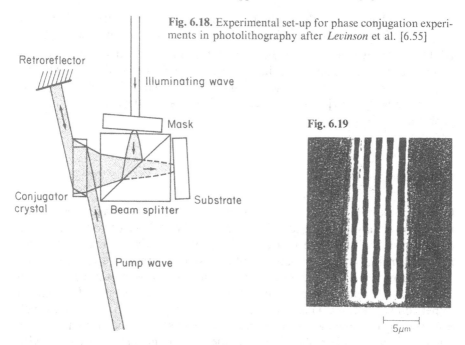

Fig. 6.18. Experimental set-up for phase conjugation experiments in photolithography after *Levinson* et al. [6.55]

Retroreflector

Illuminating wave

Mask

Fig. 6.19

Conjugator crystal

Beam splitter

Substrate

Pump wave

5 μm

Fig. 6.19. Developed pattern of 0.75 μm wide stripes with 0.5 μm separation demonstrating that submicrometer photolithography can be achieved by lensless imaging with phase conjugation [6.55]

on a 400 Å opaque chromium film deposited on a flat-glass substrate. The object wave partly reflected from the beam-splitter cube is mixed with the pump beams in the photorefractive crystal, where the phase-conjugate wave is generated. The retroreflected conjugate wave then forms an image on the surface of the photoresist-coated substrate. For the experiments reported in [6.55], the numerical aperture was calculated as N.A. = 0.48 which implies a resolution limit [6.57] of

$$\Lambda \simeq \frac{2}{\pi} \frac{\lambda}{\text{N.A.}} = 0.58 \,\mu\text{m}$$

and a depth of focus of 0.47 μm. The latter alignment was achieved by Michelson interferometry of white light reflected from the mask and the substrate [6.55]. The theoretical limit of N.A. was 0.65, i.e. very high compared to conventional projection systems. Figure 6.19, which shows a scanning-electron micrograph picture of the developed substrate, illustrates that submicrometer resolution can be achieved by lensless imaging through phase conjugation. Since most of the optical aberrations are corrected in the projection process using phase conjugation and all surfaces in the conjugator can be plane, the system described in [6.55] achieves its full resolution over a large usable field. The prime technical difficulty in the experiment reported in [6.55] was the low brightness of the projected image, which was due to the small reflectivity of the phase-conjugate material being used in this early experiment ($LiNbO_3$). The use of high-reflectivity photorefractive materials ($BaTiO_3 : R = 10000\%$ [6.58], $Ba_xSr_{1-x}Nb_2O_6$ [6.59] $R > 100\%$, $KNbO_3$ $R > 25\%$ [6.60]) would overcome this problem.

6.10.2 Adaptive Optics

Optical phase conjugation can also be used for power and image transmission through phase aberrating media. The principle of wavefront correction has been described in Sect. 6.9. The first example of phase conjugation with dynamic recording media applied in adaptive optics has to do with the transmission of an image through a nonperfect optical fiber [6.61]. A problem in transmitting an image along an optical fiber and recovering it at the other end arises from modal dispersion in the fiber. This causes a scrambling of spatial information, which can seriously degrade the image. By sending the image through a length of fiber into a conjugator and finally through another fiber identical to the first one, with the conjugator at the midpoint of the path, one could offset the effects of modal dispersion and recover the original image. *Alley* and *Dunning* have demonstrated this effect using a photorefractive $BaTiO_3$ crystal as the phase conjugate mirror, with a cw argon laser as the light source [6.62].

Another example of potential adaptive-optical application using laser-induced gratings is the problem of maintaining a diffraction-limited irradiance of a high-power laser system to be focussed on a given target.

Figure 6.20 shows a comparison of a conventional high-power laser-amplifier system proposed for laser fusion and a system employing a phase-conjugate mirror [6.73]. The conventional system (Fig. 6.20a) uses a long chain of laser amplifiers that may gradually introduce distortions in the beam arriving at the fusion target. In the phase-conjugate laser system (Fig. 6.20b) a spatially broad low-intensity laser illuminates the pellet. A small fraction of this illumination is reflected off this fusion target into the solid angle of the focussing optics and is amplified, phase-conjugate reflected and further amplified on its return. Because the phase-conjugate beam exactly retraces its path, the amplified beam is automatically focussed back to the fusion target. In addition, any phase distortions imparted on the beam by the complex amplification system will also be compensated on the return path.

The use of a phase-conjugate mirror could not only compensate for such phase aberrations, but can also essentially act as a real-time pointer and tracker that can follow the target. (This, of course, depend on the target displacement during a round-trip optical propagation time.) The result is to maintain a diffraction-limited spot on the target, even under dynamical conditions.

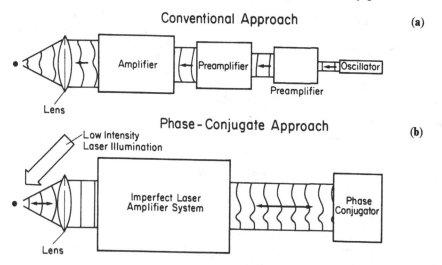

Fig. 6.20. (a) Conventional high-power laser system used, e.g., for laser fusion. (b) Optical phase conjugation applied to high-power lasers for laser fusion. Scattered light from the target is amplified by the laser system, phase-conjugated, focussed aberration-free and amplified onto the target [6.73]

6.11 Self-Pumped Phase Conjugators

Phase conjugation of optical waves by laser-induced gratings is most commonly achieved by four-wave mixing (Sect. 6.9) or stimulated scattering [6.49, 50]. The experimental configuration for the stimulated-scattering techniques is most simple (Fig. 6.21a), because the phase-conjugate wave is generated by the incident object wave interacting with acoustic phonons or density fluctuations in liquid cells resulting in stimulated Brillouin or Rayleigh scattering, see Sect. 1.2.2. Since the nonlinearities of this process are rather small, one obtains large intensities of the phase-conjugate wave only at high incident power, i.e. pulsed laser sources are required. The traditional four-wave mixing scheme (Fig. 6.21b) can, however, also be simplified by using special recording configurations (Fig. 6.21c–f). These schemes can yield high phase-conjugate reflectivities even at low cw laser powers, e.g., in photorefractive gratings.

Jensen and *Hellwarth* [6.66] have shown that only one pumping beam is required if the four-wave mixing takes place in an optical waveguide or fiber, because then the incident wave can also serve as the second pumping beam.

White et al. [6.67] have demonstrated that phase conjugators which use four-wave mixing can be self-pumped if a pair of mirrors are aligned, as shown in Fig. 6.21d to form a resonator cavity containing a photorefractive BaTiO$_3$ crystal. Both pumping beams are generated from the input wave by two-wave mixing of the input wave with scattered light propagating towards the mirrors.

OPTICAL PHASE COJUGATION

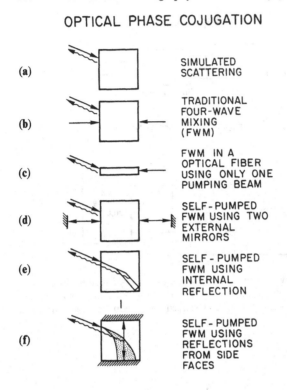

(a) SIMULATED SCATTERING

(b) TRADITIONAL FOUR-WAVE MIXING (FWM)

(c) FWM IN A OPTICAL FIBER USING ONLY ONE PUMPING BEAM

(d) SELF-PUMPED FWM USING TWO EXTERNAL MIRRORS

(e) SELF-PUMPED FWM USING INTERNAL REFLECTION

(f) SELF-PUMPED FWM USING REFLECTIONS FROM SIDE FACES

Fig. 6.21a–f. Pictorial summary of optical phase-conjugation techniques. (The phase-conjugate output is represented by a wavy arrow) [6.64, 65]

Cronin-Golomb et al. [6.68] have shown, that one of the mirrors can be removed and the phase conjugator still operates if it is started with a weak beam.

Recently two self-pumped phase conjugator configurations that need no external mirrors and which are both self-starting have been demonstrated with BaTiO$_3$ crystals (Fig. 6.21e and f) [6.64, 69]. The pumping beams are self-generated from the incident wave and are internally reflected at the crystal faces. Since the pumping beams do not leave the crystal, the device is compact, relatively insensitive to vibration, and completely self-aligning. The quality of phase conjugation by the above device was demonstrated by focussing complicated images into the crystals and observing the faithfulness of reproduction even though the image-bearing input and output beams passed through a severe phase aberrator.

Figure 6.22 shows a photomicrograph of self-pumped phase conjugation using one crystal edge as the internal reflector. A beam entering the crystal from the left with extraordinary polarization suffers asymmetric self-defocussing (Fig. 6.22a–c) in the plane formed by the incident beam and the c-axis of the crystal, creating a fan of light that illuminates one of the edges of the crystal [6.65, 70]. This edge acts as a two-dimensional corner-cube reflector and, by means of two internal reflections, it directs the fan of light back toward the incident beam. Figure 6.22b shows that the fan actually collapses into at least two narrow beams. It is speculated that each beam is composed of a pair of

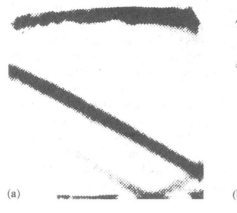

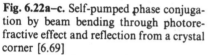

(a)

(b)

Fig. 6.22a–c. Self-pumped phase conjugation by beam bending through photorefractive effect and reflection from a crystal
corner [6.69]
(**a**): incident beam passing the BaTiO$_3$
crystal before the photorefrative grating
was built up; (**b**): beam bending and reflection from crystal corner after photorefractive grating was built up; (**c**) four-wave
mixing configuration of (**b**) (schematic)

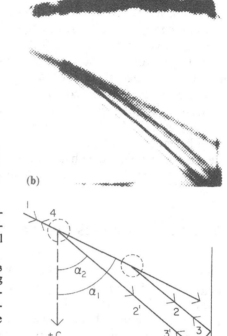

(c)

counterpropagating beams, so there are two counterpropagating loops of light
in the crystal, see Fig. 6.22c. Each pair of counterpropagating beams mixes, by
the photorefractive effect, with the incident beam (1) to create the phase-
conjugate signal beam (4). The phase-conjugate beam leaves the crystal exactly

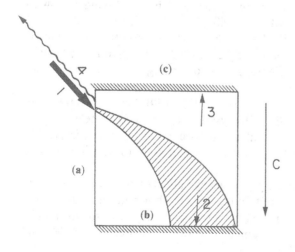

Fig. 6.23. Self-pumped phase
conjugation by beam fanning
through photorefractive effect
and reflection from a pair of
coated crystal faces [6.64]

back along the direction of the incident beam. Phase-conjugate reflectivities of up to 72 % have been measured in this configuration [6.64].

Another configuration for self-pumped phase-conjugation that is completely self-contained is shown in Fig. 6.23 [6.64]. The object beam incident through face a is fanned towards the face b which is covered by a reflecting layer and reflects the diffracted wave towards the face c which is also covered by a reflecting layer. Each pair of counterpropagating beams between the mirrors b and c mixes with the incident beam by the photorefractive effect to create a phase-conjugate beam which leaves the crystal back along the direction of the incident beam. Depending on the orientation of the crystal with respect to the incident beam direction, the phase conjugate beam can have a stable reflectivity, but it may also self-pulsate or oscillate in a chaotic manner [6.64].

6.12 Optical Resonators with Phase-Conjugate Mirrors

An interesting application of optical phase conjugation using laser-induced gratings is the use of a phase conjugator as an element in an optical resonator. One or both of the conventional mirrors can be replaced by a phase conjugating mirror. The properties of such devices have been reported in several papers [6.47, 67, 68, 71, 72]. Such resonators allow the real-time correction of distortions due to defects in laser rods, turbulence in gaseous or liquid gain media, imperfect optics, thermal and nonlinear effects which can severely limit the performance of conventional normal mirror resonators. These distortions may be particularly severe in high-power laser systems.

Phase-conjugate mirrors based on degenerate four-wave mixing generally require two additional reference beams, which may complicate the build-up of resonators with phase-conjugate mirrors. It has been shown in Sect. 6.11 that there exist mechanisms for self-pumping phase-conjugate mirrors by self-diffracting part of the wave to be conjugated and by employing four-wave mixing of the incident beam with the self-diffracted and retroreflected self-diffracted beam. In Sect. 6.11 it has been shown that in photorefractive media the phase-conjugated wave can be efficiently generated in a self-contained crystal cube. Figure 6.24 shows the different resonator arrangements using photorefractive crystals.

A phase-conjugate resonator as shown in Fig. 6.24b has been built-up by *Feinberg* [6.58]. He employed a BaTiO$_3$ crystal as a phase-conjugate mirror and an ordinary piece of silvered metal as the second resonator mirror. The resonator occurred after a few tens of seconds required for the grating build-up in the BaTiO$_3$ crystal.

The pumping beams can be self-generated by optical two-wave interaction in the photorefractive crystal [6.67, 68]. In the systems shown in Fig. 6.24c and d, lasing with the phase-conjugate resonator is not self-starting. Lasing is initially induced at the high-gain line of an argon laser at 488 nm between mirror M1 and

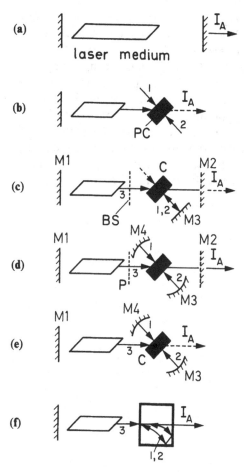

Fig. 6.24a–f. Optical Resonators: (a) Conventional two mirror system. (b) System with one phase conjugate mirror and one conventional one. The phase-conjugate mirror has to be pumped externally by two antiparallel pump beams. (c), (d) Self-pumped phase-conjugate resonator which is not self-starting. (e) Self-pumped phase-conjugate resonator self-starting using two-wave mixing. (f) Self-pumped phase conjugate resonator self-starting using beam bending

a beam splitter BS. Light transmitted through the beam splitter pumps the crystal C which is placed in the M3–M4 resonator of Fig. 6.24d or half-resonator C–M3. Gain through two-wave mixing (Sect. 4.5 and 6.6) causes an oscillation to build up in the resonator C–M_3 or M3–M4. The reflecting mirror M2 can be used to assist in the build-up. With oscillation established, the beam splitter and the retroreflecting mirror M2 can be removed (Fig. 6.24e). The starting configuration described above is required since the coherence of the fluorescence is insufficient to allow the grating formation in the crystal. The distortion correction capability of the arrangement shown in Fig. 6.24d was demonstrated by inserting an etched glass plate (P) into the normal laser cavity. This distortion destroyed the mode and lowered the output power from 2 W to 1 mW at 38 A tube current [6.67]. With the phase conjugate mirror configuration, the mode shape is restored and the power output increased to 500 mW.

Other resonator configurations have been developed which are self-pumped and self-starting (Figs. 6.24e and f). The configuration shown in Fig. 6.24f uses a self-pumped phase conjugator (Sect. 6.11) requiring no additional mirrors for generating the reference beams [6.69, 74]. Another version used a self-pumped LiNbO$_3$ crystal as the end mirror of a Cu vapor laser. The spontaneous emission from the laser is sufficient to initiate operation [6.75].

A peculiarity of laser-resonators with self-pumped phase-conjugating mirrors is that the accumulated phase shift of an optical wave propagating between a phase-conjugate and an ordinary mirror is always zero. Consequently, a phase-conjugate resonator of length L can support any wavelength consistent with the bandwidth of the gain medium and the conjugator itself. Oscillation at a particular laser wavelength will not be disturbed by any variation of the cavity length, in contrast to an ordinary resonator, whose spectral output will exhibit mode hopping and frequency drift if the cavity length varies.

7. Gratings in Laser Devices and Experiments

In this chapter, some dynamic grating effects are discussed, which are relevant to lasers and other optical devices and experiments. In lasers, a standing light-wave builds up between the mirrors. The light energy density is spatially modulated and stimulated emission produces a grating-like modulation of the population of the upper laser level. These spatial holes affect laser action in various ways, as discussed in Sect. 7.1. A spatial modulation of the optical properties of lasers can be produced also by external means, e.g. optical pumping, and is used as distributed feedback instead of laser mirrors (Sect. 7.2). Dynamic gratings are also applied to deflect, modulate and filter optical beams (Sects. 7.3–5) and to investigate coherence properties (Sect. 7.6). The coherent coupling artifact or coherence peak in sampling experiments with ultrashort light pulses (Sect. 7.7) is another example for the occurence of dynamic gratings in laser technology. It is expected that dynamic gratings will be increasingly used in future photonic systems and quantum electronics extending the range of applications described in the following.

7.1 Spatial Hole-Burning in Lasers

During laser action, a standing light-wave builds up in the cavity due to multiple reflections of the laser beam between the parallel mirrors. At the nodes of the standing wave, the light energy density is small whereas at the antinodes the density is maximum, as shown in Fig. 7.1. The population density N_2 in the upper laser level is depleted by stimulated emission preferentially at the antinodes of the light wave. The spatially inhomogeneous depletion of N_2 is called spatial hole-burning and the minima of N_2 are called spatial holes.

Optical gain, producing laser action, depends both on the population N_2 of the upper and on the population N_1 of the lower laser level. N_1 is also spatially modulated due to the optical transitions from the upper laser level. To describe the influence of both populations, N_1 and N_2, the inversion $N_2 - N_1$ is often used. Hole-burning is producing spatial holes also in the inversion.

Figure 7.1 shows the distribution of the field strength and the spatial holes in a laser cavity with two mirrors M1, M2 for a single longitudinal mode oscillating. The situation becomes more complicated if several longitudinal modes are

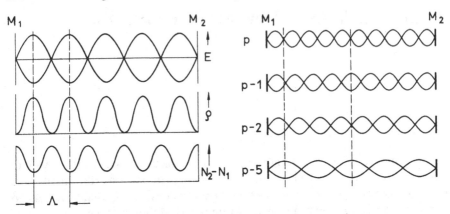

Fig. 7.1. Distributions of field strength E, energy density and population inversion $(N_2 - N_1)$ in a laser cavity with mirrors M_1, M_2

Fig. 7.2. Field strength distribution of longitudinal modes with different numbers $p, p-1 \ldots$ of maxima along the laser axis

present, as shown in Fig. 7.2. Each mode has a different field strength distribution. The nodes of the different modes usually do not coincide, so that each mode produces a different spatial distribution of holes which overlap in multimode laser operation.

Spatial hole-burning affects laser operation in the following ways. First, the spatially inhomogeneous inversion results in different total optical gains for the different modes [7.1–5]. In the case of a spatially homogeneous inversion, the gain for all longitudinal modes would be equal if the frequency dependence of the gain is neglected. Secondly, the spatial holes produce an optical grating and diffract a light wave into a direction antiparallel to the incoming direction of propagation [7.6–10]. The diffracted wave interferes with the counter-propagating light wave in the cavity. This interference is destructive if spatial hole-burning produces an amplitude grating, which is usually assumed in a first approximation. The destructive interference corresponds to an additional loss for the laser mode, or can be considered as a mechanism contributing to gain saturation. In ring lasers, the spatial hole grating couples the oppositely travelling waves due to diffraction. Because the diffracted wave is out of phase with the other incoming wave, the modes tend to suppress each other so that bistable unidirectional ring-laser operation may result [7.6].

Spatial hole-burning [7.1] was considered first for the ruby laser to explain longitudinal multimode operation and the statistical temporal spiking behaviour of the output power in pulsed ruby and other solid-state lasers, e. g. neodymium lasers.

The ruby-laser emission line is homogeneously broadened, which means that monochromatic light interacts with all atoms in the ground or excited state, thus resulting in absorbing or amplifying transitions. In contrast, a spectral line is inhomogeneously broadened, when it is possible to assign to different transitions frequencies within the line width, different groups of atoms. For example, in a gas with Doppler broadening, different transition frequencies correspond to atoms with different thermal velocities. Transitions at the center frequency are due to atoms with a zero-velocity component in the propagation direction of the absorbed or emitted light wave.

In a laser with an inhomogeneously broadened transition, like in a gas laser, usually several longitudinal modes oscillate, because each longitudinal mode has a slightly different frequency and therefore its own reservoir of inverted atoms. In an homogeneously broadened laser, the central mode has the largest gain, reaches threshold first and depletes the total inversion so that the other modes should never reach threshold. The laser thus should operate in a single longitudinal mode.

Experimentally, it has been observed that solid-state lasers operate in a large number of longitudinal modes despite homogeneous line-broadening. For an explanation, the spatial distribution of the inversion produced by the first mode reaching threshold should be considered (Fig. 7.1). The first mode depletes the inversion in a spatially inhomogeneous form. A second mode which has a different energy-density distribution (Fig. 7.2) may therefore find enough inversion and reach threshold if the pump power is high enough. In a similar way, other modes may also start to oscillate. Multimode oscillation leads to irregular spiking in pulsed systems. If a laser is switched on, relaxation oscillations of the mode amplitudes occur until steady state is reached. The relaxation oscillations are nonsinusoidal but exhibit rather sharp spikes, because of the nonlinear interaction of the laser light with the atoms in the laser cavity. Single-mode operation leads to regular spiking. The spikes corresponding to different modes have different repetition periods and are strongly coupled. In addition to the longitudinal modes, transversal modes must also be considered. The sum of the mode spike trains results in a chaotic temporal modulation of the laser output.

The spectral output and spiking behaviour of solid-state lasers, like ruby and neodymium lasers, has been investigated by various authors [7.1–6] by considering spatial hole-burning. Qualitative agreement with the experimental observations was obtained. A direct experimental proof of the spatial holes in ruby laser has been performed by Bragg diffraction of an argon laser beam, as discussed already in Sect. 1.2.1.

In gas lasers, spatial hole-burning usually is unimportant because the inversion grating is washed out by the rapid thermal movement of the gas atoms or molecules.

In semiconductor (e.g., AlGaAs, InGaAsP) lasers, spatial hole-burning by longitudinal modes has not yet attracted much attention [7.4, 11]. Rapid carrier diffusion seems to prevent the build-up of an inversion grating by longitudinal modes [7.4]. The period of such a grating would be only $\Lambda = \lambda/2n \approx 100 \ldots 200$ nm resulting in a diffusive grating decay time of $\tau_D = \Lambda^2/4\pi^2 D_a \approx 10^{-12}$ s if a diffusion constant $D_a \approx 10 \text{ cm}^2/\text{s}$ is assumed (Sect. 5.5). Because of homogeneous broadening of the laser line and negligible spatial hole-burning, transverse-mode stabilized InGaAsP lasers tend to operate in a single longitudinal mode [7.13].

Transversal modes also deplete the inversion inhomogeneously in a direction perpendicular to the laser axis. In contrast to hole-burning by longitudinal modes, the distribution of the inversion is not spatially periodic but reflects the transversal mode structure, e.g. a Gaussian distribution for the fundamental mode. The dimensions of this distribution are typically of the order of several µm so that diffusion is less important. Transverse hole-burning in semiconductor lasers influences the transverse mode structure [7.12, 13].

In cw dye lasers, spatial hole-burning leads to multilongitudinal-mode operation similarly as in solid-state lasers. As has been discussed briefly in Sect. 1.2.1, unidirectional traveling wave ring lasers can be used to avoid spatial hole-burning and to achieve high-power single-mode operation. For this purpose it is, however, also possible to use linear cavities with filters and other mode selectors. Low insertion losses of these systems are necessary and can be achieved with 3-mirror reflectors [7.14, 15] or Michelson selectors [7.16, 17].

In mode-locked dye lasers producing femtosecond pulses, spatial hole-burning seems to play an important role in coupling the pulses propagating in the cavity and in determining the final pulse width [7.18, 19].

Closely related to spatial hole-burning is the production of amplitude and phase gratings in saturable absorbers inside laser cavities. It must be expected that these gratings strongly affect the performance of passively Q-switched and mode-locked lasers [7.20]. It has been demonstrated experimentally that pure organic solvents could themselves partially Q-switch ruby and neodymium lasers in the absence of a saturable dye [7.21]. Q-switching arises from the enhancement of the reflectivity of the liquid, during the evolution of the laser pulse, through the formation of a periodic refractive index modulation in the liquid by the action of the standing light-waves in the cavity. The refractive-index periodicity acts as a reflector providing a dynamic increase of the cavity Q-factor. The optical Kerr effect (Sect. 3.9) and thermal effects (Sect. 3.7) by one- or two-photon absorption were considered as mechanisms producing refractive index changes in the liquids used like chloronaphtalene, methanol, aceton and water.

7.2 Distributed-Feedback Lasers

Distributed-feedback lasers usually consist of an active medium and dielectric mirrors which are formed by a sequence of alternating thin films with high and low refractive index. Volume gratings with a harmonic distribution (period Λ) of the refractive index can also be used as mirrors. Light with a vacuum wavelength λ propagating in a medium with a mean refractive index n is reflected if the Bragg condition, equation (4.27), is fulfilled. For a beam incident normal to the grating planes this condition may be written as $\lambda/2n = \Lambda$.

Kogelnik and *Shank* [7.22] have shown that a periodic structure in a laser-active material can be used simultaneously for confinement and amplification of radiation. The feedback provided by the laser mirrors is usually distributed over the whole laser. Distributed feedback is increasingly used for injection and dye lasers. Permanent periodic structures produced, e.g., by etching are often used.

It is possible to use also transient gratings for feedback. These are induced by interference of two laser beams and have the advantage of tunability by changing the grating period. The first laser of this type used Rhodamine 6G dissolved in alcohol mixtures [7.23]. The grating was written with a frequency-doubled ruby laser which simultaneously pumped the dye. The laser was tunable from 572 to 536 nm by changing the angle between the interfering pump beams.

At high pump powers, higher-order Bragg reflections are possible, resulting in a generalized resonance condition [7.24]:

$$\lambda^{(m)}/2n = \Lambda/m, \quad m = 1, 2, 3 \ldots .$$

Transient distributed-feedback lasers have been constructed also in thin films [7.25] and other solid materials [7.34] doped with laser dyes.

It was reported that distributed feedback in Rhodamine 6G solutions has been obtained not only by periodic but also by homogeneous pumping. The amplified luminescence is reflected from a single mirror and interferes with the incident luminescence resulting in a grating [7.27]. In later work with a different dye solution, the wavelength region of distributed-feedback lasers was extended to cover the wavelength region from 400 to 900 nm [7.26, 28–32].

Population density and temperature gratings have been considered as feedback mechanisms [7.31]. These gratings have been called amplitude and phase gratings although a population density change results not only in an absorption change but also gives a wavelength-dependent index change, as outlined in Sect. 3.4. An extensive theoretical discussion of laser performance in both index and gain grating configuration has been given in [7.33].

Distributed-feedback dye lasers (DFDL) have been pumped with nitrogen and excimer lasers or frequency-doubled or tripled ruby and neodymium lasers. The excitation pulse width typically amounts to 1–10 ns. The DFDL emission then consists of a train of picosecond pulses which are due to relaxation oscillations [7.35, 36]. The theoretical description [7.37, 38] is based on rate equations for the excited-state population and DFDL photon density. A similar description is used to explain the spiking behavior of solid-state lasers. Single pulses are generated if the pump energy is in the range of 1–1.2 times the threshold value. However, even when pumped well above threshold, the first pulse of the pulse train can be isolated by quenching the further laser emission. An additional quenching laser has been used for this purpose but a simplified scheme seems possible [7.40]. Pumping with 2.5 ns long pump pulses from a nitrogen laser, one can achieve a DFDL pulse width of 17 ps at 380 nm.

The advantages of the DFDL as compared to other dye lasers are simplicity, reliability and easy tunability achieved by changing the angle between the excitation beams. Due to these advantages, the DFDL may be useful in practice especially for picosecond pulse generation.

7.3 Optical Light Deflection and Modulation

Laser-induced dynamic gratings can be used for deflecting or modulating other light waves. An opto-optical deflector or modulator operates in a similar way as the well-known acousto-optic devices, where light diffraction occurs from the periodic refractive index modulation introduced into the acousto-optic material by a travelling acoustic wave [7.41]. The speed of such devices is limited by the sound velocity. It seems straightforward to replace the acoustic waves by a laser-induced transient grating to obtain much faster response times. A further advantage of such devices would be that no electrically driven transducers are required, so that applications in all-optical systems seem possible, e.g. switching in fiber-optic communication systems.

Opto-optical deflection devices for discrete angles can be realized by a simple transient grating arrangement discussed throughout this volume. The problems of continous light scanning have been discussed in [7.42, 43]. Continous deflection can be done with a wavelength tunable laser which produces gratings with changing period Λ, resulting in different diffraction angles. Wavelength modulation could be easy with semiconductor lasers by changing the drive current.

Figure 7.3 shows that a small deviation $\Delta\theta$ from the Bragg diffraction angle given in (2.46) can result in a large change of the diffraction efficiency. For continuous scanning this means that the grating has to be tilted in combination with wavelength scanning, so that the grating planes are always parallel to the bisectrix of the incident and diffracted beams. This can be done by passive optical devices consisting of static diffraction gratings and lenses (Figs. 7.4, 5).

Opto-optical light deflection has been demonstrated experimentally by using the fixed argon laser lines between 477 and 514 nm to write photorefractive gratings into $LiNbO_3$ and $Bi_{12}SiO_{20}$ crystals [7.42, 43]. Deflection of a He-Ne laser at 633 nm was demonstrated with constant diffraction efficiency over deviation angles up to about 30°.

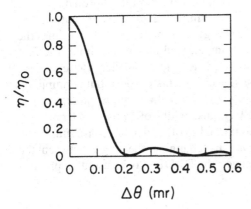

Fig. 7.3. Angular variation of the diffraction efficiency for a 1 cm thick laser induced grating in $LiNbO_3$ [7.42]

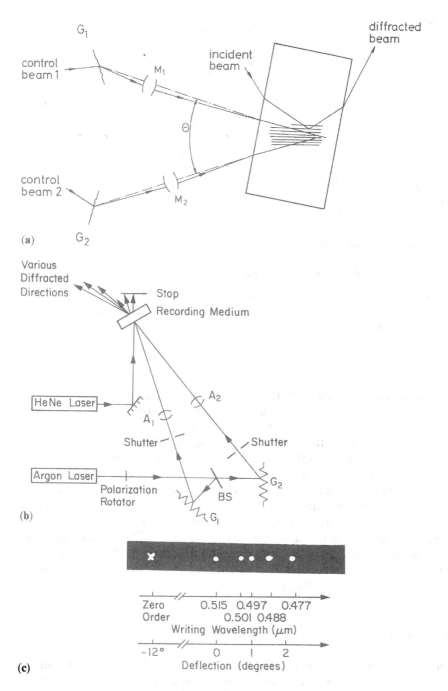

Fig. 7.4a–c. Optical deflection of a HeNe laser beam into different direction. The grating is written by an argon laser at various wavelengths. (**a**) Enlarged sketch of the interaction region. Fringe tilt is achieved by frequency dependent diffraction of the control beams at gratings G_1, G_2. (**b**) Experimental arrangement. (**c**) Deflection angles obtained with various writing wavelength, after *Sincerbox, Rosen* [7.42]

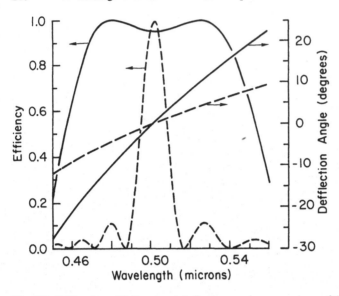

Fig. 7.5. Diffraction efficiency and deflection angle comparison of fringe tilting (——) and static fringe (– – –) light deflectors [7.42]. Deflection angle increases monotonously with writing wavelength

To achieve a simultaneous change in both the period Λ and the orientation of the grating while varying the wavelength of the two writing beams, *Sincerbox* and *Roosen* [7.42] have conceived a passive technique that operates in real time over a wide angular deflection range. This is shown in Fig. 7.4 where the incoming beam is split into the two identical writing beams by a beam splitter each of which is incident on a dispersive medium, in this case the diffraction gratings G_1 and G_2. Changing the wavelength of the incident beam causes a change in the direction of the diffracted beams and consequently a change in the direction of the beams incident on the recording medium. Using gratings with different spatial frequencies, the change in each beam was different and therefore a simultaneous change in both the spacing and the direction of the interference grating is obtained. The resulting increase of the high efficiency deflection angle range is demonstrated by Fig. 7.5.

7.3.1 Deflection by Nonlinear Gratings

Other methods to diffract a laser beam at one wavelength by means of a laser-induced grating recorded at another wavelength were proposed by *Petrov* et al. [7.45, 46], and *Huignard* and *Ledu* [7.44]. They are based on anisotropic diffraction in birefringent crystals [7.45] or on nonlinear mixing of two gratings having independent spatial frequencies q_1 and q_2 [7.44].

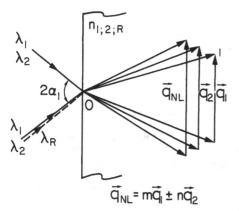

Fig. 7.6. Wave-vector diagram illustrating the collinear Bragg diffraction. Grating wave vector q_{NL} is induced by nonlinear mixing of grating wave vectors q_1 and q_2 [7.44]

In the method proposed in [7.44] the two gratings q_1 and q_2 are produced by a proper choice of collinear beams at wavelengths λ_1 and λ_2. For readout at λ_r, exact collinearity of writing (λ_1, λ_2) and reconstructed (λ_r) light beams is obtained.

Exposing the photosensitive material with two collinear intensity grating vectors q_1 and q_2, as shown in Fig. 7.6, can generate refractive-index grating vectors $q_{NL} = mq_1 + nq_2$ through nonlinear mixing. The Bragg condition for a collinear readout wavelength λ_r is satisfied if

$$\frac{1}{\lambda_r} = \frac{m}{\lambda_1} \pm \frac{n}{\lambda_2} , \quad \text{where} \quad m, n = 1, 2, 3 \dots . \tag{7.1}$$

In photorefractive $Bi_{12}SiO_{20}$ used in the experiment described in [7.44] the nonlinear mixing of wavevectors q_1 and q_2 arises from the nonlinear expression for the photoinduced refractive index change given by, see (3.85),

$$\Delta n(x) = \frac{A}{1 + m_1(\cos q_1 x + \cos q_2 x)} , \tag{7.2}$$

where

$$A = \frac{n^3}{2} r E_0 (1 - m_1^2)^{1/2} .$$

Expanding this relation into a Fourier series gives

$$\Delta n(x) = \sum_{n=0}^{\infty} \sum_{m=0}^{\infty} \alpha_{mn} [\cos(mq_1 + nq_2)x$$

$$+ \cos(mq_2 - nq_1)x] , \tag{7.3}$$

with $\alpha_{m,n}$ as the amplitudes of the nonlinear grating components.

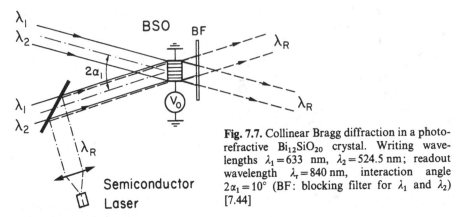

Fig. 7.7. Collinear Bragg diffraction in a photo-refractive $Bi_{12}SiO_{20}$ crystal. Writing wave-lengths $\lambda_1 = 633$ nm, $\lambda_2 = 524.5$ nm; readout wavelength $\lambda_r = 840$ nm, interaction angle $2\alpha_1 = 10°$ (BF: blocking filter for λ_1 and λ_2) [7.44]

Expression (7.1) which describes the matching condition of the photoinduced nonlinear grating produced by λ_1 and λ_2, for the read-out wavelength λ_r can be fulfilled for different sets of (m, n).

For $n = 0$, $m \geq 2$, (7.1) expresses the collinear Bragg diffraction at a wavelength that is shorter than the writing wavelengths. These wavelengths would be strongly absorbed by the crystal (uv spectral range), and erasure would occur during readout. This type of nonlinearity permits the observation of second order diffraction ($m = 2$) of a photographic hologram with a readout beam collinear to the reference but whose frequency is twice the recording frequency [7.47].

By a proper choice of parameters m and n in relation (7.1), a near-ir readout wavelength can be obtained while keeping the incident wavelengths λ_1, λ_2 in the spectral-sensitivity range of the crystal. The wave-vector diagram corresponding to such a nonlinear interaction is given in Fig. 7.6. This geometrical construction illustrates that the Bragg condition is fulfilled, thus permitting a exact collinearity of the incident recording $(\lambda_1\lambda_2)$ and diffracted (λ_r) beams.

Photoinduced collinear Bragg diffraction with diffraction efficiency of up to 5×10^{-4} was observed in [7.44] (Fig. 7.7) for $\lambda_r = 840$ nm ($Ga_{1-x}Al_xAs$ laser diode) with $\lambda_1 = 633$ nm, $\lambda_2 = 514$ nm and $(m = 2, n = 1)$. The condition (7.1) would have required $\lambda_r = 824$ nm for the above parameters neglecting material dispersion. This was not the emission wavelength of the diode laser and therefore the diffraction efficiency was reduced.

To apply the demonstrated devices in optical communication systems, it seems necessary to extend the principles to optical waveguides and integrated structures.

7.3.2 Optical Gating

A dynamic grating can also be used as an optical gate [7.48, 49]. In the simplest version, the grating is produced by a short probe pulse in a sufficiently fast nonlinear optical material [7.49]. An additional signal pulse is diffracted only as long as the gate pulse is present. The idea has been tested, e. g., with a CS_2 Kerr liquid and a mode-locked Nd:glass laser, resulting in a time resolution of less than 10 ps. Compared to the usual optical Kerr shutters using laser-induced birefringence, the grating shutter requires no analyzing element such as a polarizer because the propagation directions of the gated and incident light are different.

It has been pointed out [7.48] that fast optical gating can be obtained also with gratings in materials with slow response. The wavefront conjugation arrangement described in Fig. 4.10 is applied for this purpose. Two counter-propagating pump waves are used. Interference of one pump wave with the signal wave produces a grating which diffracts the other pump wave into a direction opposite to the signal wave. The amplitude of this backward wave is proportional to the instantaneous amplitude of the signal wave if it is assumed that the pump pulsewidth is short compared to the signal duration. The delay time of the second pump beam can be varied and the amplitude of the signal wave is thereby probed or gated at different times. The backward wave thus corresponds to the gated wave. The time resolution, or gate width, is given by the pump pulse width.

7.3.3 Pulse Compression

If a dynamic grating in a sufficiently fast material is excited and probed with the same laser pulse, a pulse compression results due to optical gating, as described above. A simple experimental realization is obtained by self-diffraction or degenerate four-wave mixing (DFWM). Pulse compression by intracavity DFWM of picosecond pulses in CS_2 has been suggested in [7.50]. DFWM of ultrashort pulses from a Rhodamine 6G laser with pulse energies of 0.5 mJ at 620 nm yielded high-power pulses of less than 40 fs duration [7.51].

7.4 Optical Diodes and One-Way Viewing Windows

Laser-induced dynamic gratings in media with non-local response (e.g., photorefractive materials), where the photoinduced gratings may be phase-shifted with respect to the light interference fringes, can lead to a non reciprocal energy-exchange between writing beams (Sect. 4.5). As shown in Sect. 6.6 this coupling between recording beams can be used for coherent amplification of time-varying optical images. Such beam-coupling effects exist also if a grating is produced by two antiparallel beams (reflection-hologram configuration). The asymmetric intensity transfer between antiparallel beams leads to a new type of unidirectional device, with the characteristics of an optical diode which transmits light preferentially from one side of the crystal.

Figure 7.8 shows the two configurations which we shall discuss in this section. The first one uses two equal incident intensities I_{10} and I_{2L} (measured

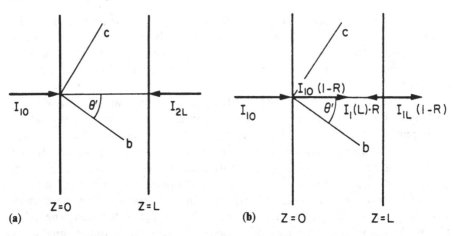

Fig. 7.8a, b. Configurations for antiparallel beam coupling experiments (b, c: crystal axes). (a) 2 incident beams, and (b) 1 incident beam interferes with a beam reflected at the exit surface $Z=L$

within the crystal). In the second configuration (Fig. 7.8b) a photoinduced grating is formed by interference of one incident beam with its Fresnel reflection at the other side of the sample.

In both cases the transfer of energy is described by the coupled wave equations [7.53] treated for a similar situation in Sect. 4.5.3:

$$\frac{dI_1}{dz} = \Gamma I_1 I_2 / (I_1 + I_2) - K I_1 \ , \tag{7.4}$$

$$\frac{dI_2}{dz} = \Gamma I_1 I_2 / (I_1 + I_2) - K I_2 \ , \tag{7.5}$$

where Γ is the intensity gain factor given by

$$\Gamma = k n_1 \sin \phi / \cos \theta \ , \tag{7.6}$$

with k being the wave vector, n_1 the amplitude of the refractive index grating induced by the space-charge fields E_{sc}, ϕ the phase-shift of the refractive index grating with respect to the interference pattern, and θ the internal angle of light propagation ($\theta = 0$ for directions perpendicular to crystal faces). In our case of diffusion recording the magnitude of ϕ is $\pi/2$ [7.52]. For the sake of simplicity let $\theta = 0$ (as shown in Fig. 7.8).

The solutions of (7.4, 5) for negligible absorption $K = 0$ are subjected to an energy conservation condition, i.e.

$$\frac{d}{dz} (I_1 - I_2) = 0 \ . \tag{7.7}$$

The solutions of (7.4, 5) at the respective exit crystal faces in the weak absorption case $(K \ll \Gamma)$ are given by

$$I_1(L) = \frac{(1+\beta)I_{10}}{\beta + \exp(-\Gamma L)} \exp(-KL) , \tag{7.8}$$

$$I_2(0) = \frac{(1+\beta)I_{2L}}{1 + \beta \exp(\Gamma L)} \exp(-KL) , \tag{7.9}$$

where $\beta = I_{10}/I_{2L}$ is the incident intensity ratio. Intensities of beams $I_1(z)$ and $I_2(z)$ and the outcoming beams $I_1(L)$ and $I_2(0)$ corresponding to (7.8, 9) are plotted in Fig. 7.9a for $\Gamma = 7 \text{ cm}^{-1}$ obtained with KNbO$_3$ crystals [7.53] (set $I_{10} = I_{2L} = 1$). One can see that the beam I_1 is amplified and I_2 is attenuated for $\Gamma > 0$.

Note that the reflection losses on the two interfaces and the absorption in the crystal have been neglected in the above treatment.

It is interesting to consider the case where only one beam enters normally to the entrance face of the photorefractive crystal and interferes with the reflected beam from the other side, creating a refractive index grating, as shown in Fig. 7.8b. For a beam which is normally incident on a crystal and reflected once with a reflectivity $R = (n-1)^2/(n+1)^2$, we can substitute $I_{10}(1-R)/RI_1(L)$ for β in (7.9) and by using the definition of the transmission factor T_1

$$T_1 = I_{1L}(1-R)/I_{10} , \tag{7.10}$$

we get

$$T_1 = 2(1-R)^2 \exp(-KL) \big| [1 - R \exp(-KL)]$$
$$+ \sqrt{[1-R\exp(-KL)]^2 + 4R\exp[-(K+\Gamma)L]} \big|^{-1} . \tag{7.11}$$

The transmission factor T_2 of the beam incident from the opposite face is simply obtained by changing the sign of Γ in (7.11)

$$T_2 = 2(1-R)^2 \exp(-KL) \big| [1 - R \exp(-KL)]$$
$$+ \sqrt{[1-R\exp(-KL)]^2 + 4R\exp[-(K-\Gamma)L]} \big|^{-1} . \tag{7.12}$$

Using the reflection R as a parameter, $(T_1 - T_2)$ can be plotted as the functions of the thickness L. Equations (7.11 and 12) reveal that the nonreciprocal light transmission through a photorefractive medium is caused by the nonlinear interaction in the medium which is characterized by the intensity gain factor Γ which depends on the $\pi/2$ phase shifted component of the photoinduced grating.

The experimental results for the wavelength dependence of the transmission factors T_1 and T_2 in Fig. 7.10 show that more than 20% asymmetry can be obtained in a $L = 2.9$ mm long crystal of KNbO$_3$ with optimized crystallographic orientation [7.53].

Fig. 7.9. (a) Intensity distribution of interacting forward (I_1) and backward waves (I_2) in a photorefractive crystal in a reflection grating geometry. (b) Intensity of outcoming beams as function of ΓL for antiparallel beam coupling with equal intensity of incident beams. (c) Transmission asymmetry ($T_1 - T_2$) versus crystal thickness ($K = 1$ cm^{-1}, $\Gamma = 5.5$ cm^{-1}). (d) Intensity distribution of incident (I_1) and beam I_2 reflected from the crystal end face at $z = L$. These beams interact in the crystal through a photorefractive grating. (——) no intensity transfer ($K = 1.5$ cm^{-1}), (— — —) intensity transfer in $+Z$ direction ($K = 1.5$ cm^{-1}; $\Gamma = +7$ cm^{-1}; $R = 0.2$), (— · — ·) intensity transfer in $-Z$ direction ($K = 1.5$ cm^{-1}; $\Gamma = -7$ cm^{-1}, $R = 0.2$) [7.53]

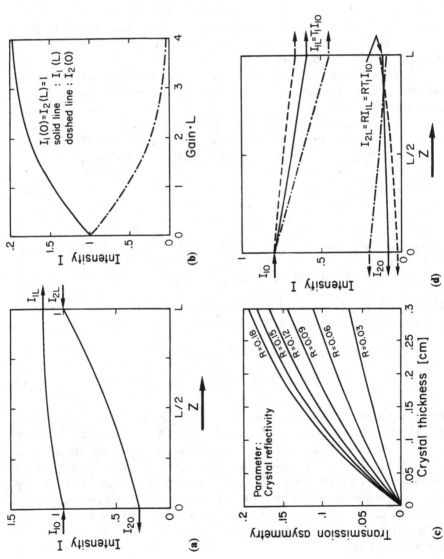

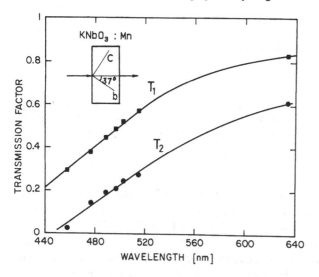

Fig. 7.10. Wavelength dependence of the transmission factor of $KNbO_3$:Mn crystal: (●) and (■) are for the two beams propagating in opposite directions [7.53]

7.5 Optical Filtering by Nearly Degenerate Four-Wave Mixing

Permanent gratings are used in many spectroscopic schemes to obtain wavelength dispersion and optical filtering. These are mostly surface gratings but volume gratings are also applied in holography to obtain wavefront storage and wavelength selectivity. It seems straightforward to use laser-induced dynamic gratings for spectroscopic applications to achieve wavelength and time selection of a probing light beam, too.

The optical filtering technique via nearly degenerate four-wave mixing [7.56–62] described here does not exploit the usual wavelength dispersion of a grating. This technique is based on the frequency response of four-wave mixing. The mixing geometry shown in Fig. 7.11 has been used in theoretical and experimental investigations. The mixing is usually described by applying nonlinear optical methods, but here we shall use the grating picture to discuss the principle of the method.

The signal wave with frequency ω_4 and the pump wave with frequency ω, which determines the center frequency of the filter, are incident on the nonlinear material. A moving grating (Sect. 4.7) is produced in the material by interference. The grating amplitude decreases with the frequency difference $\omega - \omega_4$. If $\omega - \omega_4$ becomes large compared to $1/T$, the inverse relaxation time of the material, the grating is not built up at all. The grating is probed by the counter-propagating pump wave, thus yielding a diffracted wave having maximum power at $\omega = \omega_4$. The frequency width of such a filter is $\Delta\omega \approx 1/T$.

Optical filtering by four-wave mixing was first demonstrated in CS_2 as nonlinear material and using high-power pulsed lasers [7.59]. In later investigations [7.60, 62], sodium vapor and cw or pulsed dye lasers have been used. A

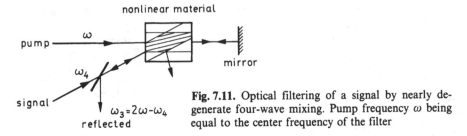

Fig. 7.11. Optical filtering of a signal by nearly degenerate four-wave mixing. Pump frequency ω being equal to the center frequency of the filter

population density grating (Sect. 3.4) is produced by tuning the lasers to the $3S_{1/2}-3P_{3/2}$ transition at 5890 Å. The frequency response of a resonant two-level system is limited by the transverse relaxation time T_2 which was estimated to be on the order of the natural lifetime $T_1 = 16$ ns. From this, a filter bandwidth of $\Delta\omega/2\pi = 13$ MHz was expected. Experimentally, a bandwidth of 40 MHz has been observed due to the frequency jitter of the dye laser. The center frequency of the filter could be tuned by detuning the pump laser from the atomic resonance. However, the efficiency of the filter decreases very quickly with detuning. Using pump beams with different frequencies, a filter can be constructed whose central bandpass is tunable over the Doppler profile with only a slight decrease in efficiency as compared with line center operation [7.61]. There should be no problem to obtain filter tunability by using nonresonant materials instead of sodium.

In summary, a four-wave mixing filter has a narrow bandwidth and is tunable. In addition, such a filter could have a large field of view because the directions of the pump and signal beam may include an arbitrary angle. For Doppler broadened resonant materials, the field of view is small [7.58]. A possible advantage is amplification of the signal wave, so that the reflection coefficient could be larger than unity. A drawback of the system is its complexity, because a pump laser has to be employed to obtain filter action, which has so far hampered widespread application.

7.6 Coherence Measurements of Laser Beams

Laser-induced transient gratings have been used to measure the temporal [7.63, 64] and spatial [7.65] coherence properties of laser beams. Only the temporal coherence is discussed here as an example.

The temporal coherence function $\Gamma(\tau)$ of a light pulse or a cw light wave is defined as the autocorrelation function of the complex field amplitude $E_1(t)$:

$$\Gamma(\tau) = \int_{-\infty}^{\infty} E_1^*(t)E_1(t+\tau)dt = \int_{-\infty}^{\infty} E_1(t)E_1^*(t-\tau)dt \ . \tag{7.13}$$

The absolute value of the coherence function can be measured by splitting the primary light beam, thus producing a two-beam interference pattern and measuring the visibility of the fringes

$$|\Gamma(\tau)/\Gamma(0)| = (I_{max} - I_{min})/(I_{max} + I_{min}) \tag{7.14}$$

as a function of the time delay τ of the two beams [7.66].

With light pulses possessing enough energy, it is possible to store the interference pattern instantaneously without further processing in a material with optical constants (index of refraction or absorption coefficient) depending on the integrated incident light power. Such a real-time storage of an interference pattern can be performed in a variety of different materials, e.g. semiconductors and dye solutions (Chap. 3). The visibility of the fringes may then be measured by diffracting a third light beam incident on the grating with a suitable time delay after excitation. The diffracted power is directly related to the visibility and has to be measured as a function of the time delay τ of the two beams producing the grating.

This new method for measurement of $|\Gamma(\tau)|$ has been demonstrated experimentally with picosecond pulses from a Nd:YAG laser [7.63, 64]. The method has been used also with Q-switched pulses from a ruby laser and should be applicable also to cw laser beams.

7.6.1 Experimental Arrangement and Theory

For measurement of the coherence function, a laser beam is split into three different directions with time delays τ and τ_0. The three beams are superimposed in a suitable nonlinear material (Fig. 7.12). The beams E_1 and E_2 intersecting at an angle θ interfere and produce a fringe pattern.

The electrical field strengths E_1, E_2 of the two beams are given by

$$E_1 = \frac{E_1(t)}{2} \exp(i k_1 \cdot r - \omega_0 t) + \text{c.c.} , \tag{7.15}$$

$$E_2 = \frac{E_1(t-\tau)}{2} \exp(i k_2 \cdot r - \omega_0 t) + \text{c.c.} . \tag{7.16}$$

Here k_1 and k_2 are the wave vectors of the two beams and r is the position vector. The complex field amplitude $E_1(t)$ describes the time dependence of the field strength with a mean frequency ω_0. The intensity I in the interference region is given by

$$I \sim \overline{(E_1 + E_2)^2} \tag{7.17}$$

$$= \frac{1}{2}\left\{|E_1(t)|^2 + |E_1(t-\tau)|^2 + E_1(t-\tau)E_1^*(t) \exp[-i(k_1 - k_2) \cdot r] + \text{c.c.}\right\} , \tag{7.18}$$

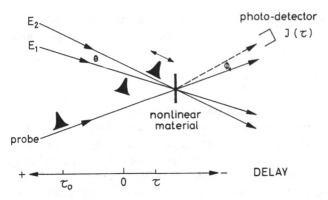

Fig. 7.12. Experimental arrangement for measurement of the coherence function $|\Gamma(\tau)|$

where the bar indicates averaging over times which are long compared to $1/\omega_0$ but short compared to the pulse coherence time. From (7.18) the spatial maximum I_{max} and minimum I_{min} of the pulse energy $\int_{-\infty}^{+\infty} I\, dt$ can be determined to prove (7.14) starting from (7.13).

The light power is absorbed in the nonlinear material and produces a spatial modulation of the optical constants which can be described by a complex refractive index. The spatial amplitude $\Delta n(\tau)$ of the refractive index is given by the spatially modulated part of the total absorbed light intensity if material relaxation is neglected:

$$\Delta n(\tau) \sim \left| \int_{-\infty}^{\infty} E_1(t-\tau)E_1^*(t)\,dt \right| = |\Gamma(\tau)| \ . \tag{7.19}$$

To determine Δn, the energy $J(\tau)$ of the diffracted probe pulse is measured. The probe pulse delay τ_0 has to be large compared to the pulse width determining the build-up of the grating. In this case

$$J(\tau) \propto |\Delta n(\infty, \tau)|^2 \propto |\Gamma(\tau)|^2 \ . \tag{7.20}$$

7.6.2 Experimental Example

In Fig. 7.13, an example for a coherence function measured for single pulses from a frequency-doubled Nd:YAG laser (530 mm) is given [7.64]. The pulse width has been determined to be 22 ps with a streak-camera. The main beam was split into three parts according to Fig. 7.12. The energies of the three pulses ranged from 0.4 to 20 µJ. The delay of the probe pulse amounted to $\tau_0 = 96$ ps. A 10^{-3} M solution of Rhodamine 6G 0.1 mm thick was used as nonlinear material. The maximum diffracted energy amounted to 0.6 nJ. The measured coherence function (Fig. 7.13) can be fitted by a symmetrical exponential function

$$|\Gamma(\tau)/\Gamma(0)| = \exp\left[-\ln 2 |\tau|/2\, t_c\right] \tag{7.21}$$

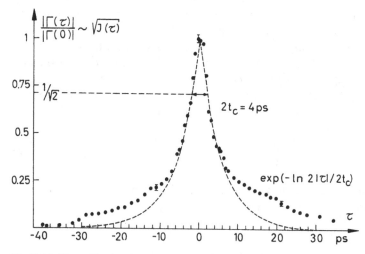

Fig. 7.13. Normalized coherence function $|\Gamma(\tau)/\Gamma(0)|$ measured in the experimental arrangement of Fig. 7.12 (see text for experimental details). The points are measured. The dashed curve corresponds to Eq. (7.21)

with a coherence time $t_c = 4$ ps. The coherence time indicates the width of the coherence function. The exact definition of the coherence time is somewhat arbitrary. It is chosen here in such a way that, for a one-sided exponential pulse, the coherence time equals the pulse width, see (7.27, 28), if no additional amplitude or phase modulation is present.

7.6.3 Comparison Between Coherence Time and Pulse-Width

The coherence time $t_c = 4$ ps is much smaller than the pulse width $t_p = 22$ ps. It is possible to explain this behaviour by assuming that a laser pulse is given by an amplitude function $E(t)$ multiplied by a phase function $f(t)$ (describing, e.g., a frequency chirp) or a random complex function $f(t)$ describing statistical amplitude or phase changes

$$E_1(t) = E(t)f(t) \ .\tag{7.22}$$

If $f(t)$ changes rapidly compared to $E(t)$, one obtains by piecewise integration

$$\Gamma(\tau) = R(\tau) \int_{-\infty}^{+\infty} E(t)E(t-\tau)dt \ ,\tag{7.23}$$

where $R(\tau)$ is the autocorrelation function of $f(t)$

$$R(\tau) = \frac{1}{\Delta t} \int_{t'}^{t'+\Delta t} f(t)f^*(t-\tau)dt \ .\tag{7.24}$$

It is assumed that Δt is short compared to the pulse width but long enough to obtain a good average which is independent of t'.

Equation (7.23) is valid not only for rapidly changing $f(t)$ but also for constant $f(t)$ and $R(\tau)$. It seems therefore meaningful to use (7.23) also as an approximation for slowly varying $f(t)$, i.e. for all possible $f(t)$.

Further evaluation of (7.23) is possible if assumptions on $E(t)$ and $f(t)$ are made. A simple example for $E(t)$ is a one-sided exponential

$$E(t) = \begin{cases} 0 & t < 0 \ , \\ E_0 \exp(-\alpha t) & t \geq 0 \ . \end{cases} \tag{7.25}$$

The width of this pulse at half-maximum power is given by $t_p = \ln 2/2\alpha$.

A typical example for $R(\tau)$ is given by

$$R(\tau) = \exp(-v|\tau|) \tag{7.26}$$

which may be obtained if $f(t)$ describes a random phase modulation of the pulse with v jumps per unit time [7.64, 68]. The coherence function of pulses described by (7.22, 5,6) is expressed by

$$\Gamma(\tau) = (E_0^2/2\alpha) \exp[-(v+\alpha)|\tau|] \ . \tag{7.27}$$

A coherence function of this type approximates the experimental observation [Fig. 7.13 and (7.21)].

The coherence time may be defined as

$$t_c = \ln 2/2(\alpha + v) \leq t_p \ , \tag{7.28}$$

so that $t_c = t_p$ for $v = 0$.

The experimental result $t_c < t_p$ indicates that the pulses contains a temporal substructure. The frequency spectrum of such a pulse is broader than is expected from the inverse pulsewidth. The exact relation between the coherence function and frequency spectrum was outlined in [7.64]. The broad pulse width and short coherence time indicate that the laser medium could support a shorter pulse. The pulse is not transform-limited or not completely mode-locked.

7.6.4 Conclusion

For picosecond pulses, $|\Gamma(\tau)|$ gives information on the pulse frequency spectrum and on the degree of mode-locking if the temporal pulse length is measured simultaneously. The transient grating method for measuring $|\Gamma(\tau)|$ needs a similar apparatus as the second-harmonic generation method [7.67] for the measurement of the duration of picosecond optical pulses. Instead of a second-harmonic crystal for the measurement of the pulse duration, a dye solution or another suitable material has to be used for the measurement of the coherence time. It seems therefore attractive to use both methods in parallel. Another motivation for the investigation of the temporal coherence of light pulses comes from the observation of coherent coupling effects in picosecond absorption measurements. The shape and width of the coherent coupling peaks is determined by the temporal coherence properties of the picosecond bleaching and probing beams (Sect. 7.7).

7.7 Coherence Peaks in Picosecond Sampling Experiments

Various laser systems have been developed that produce pulses in the picosecond [7.69] or femtosecond [7.70] range. These lasers are used for studying ultrafast transient, optical responses due to different material excitations [7.71]. The lack of sufficiently fast detectors and electronics has been overcome by using sampling techniques with a powerful pump pulse for excitation, and delayed probe pulses for detection of the material response. The optical response may be photo-induced absorption, bleaching (Fig. 7.14), dichroism or build-up of a transient diffraction grating.

A complication with the techniques arises when one tries to obtain information about relaxations occuring during the pulse duration with coherent pump and probe beams derived from the same laser. Interference effects during the overlap of the pump and probe pulses produce the so-called coherence artifact or coherence peak at zero delay time (Fig. 7.15) that makes it more difficult to extract information about the response of the material [7.72, 73].

7.7.1 Coherence Coupling in Photoinduced Absorption or Bleaching Experiments

The pump and probe technique will be first discussed in terms of photo-induced absorption or bleaching experiments, where the detected signal is linearly related to the change ΔK in the optical absorption. In transient dichroism or grating experiments, the detected signal depends quadratically on the change ΔK, Δn of the optical constants and the calculation of coherent coupling effects gets more complicated than in the linear case even though the basic mechanism is the same in both cases.

The optical absorption change ΔK produced by the pump pulse $I_1(t)$ is given by

$$\Delta K(t) = D \int_{-\infty}^{+\infty} dt' M(t-t') I_1(t') \ , \tag{7.29}$$

where $M(t)$ with $M(0) = 1$ is the impulse response of the medium to a delta-function excitation. A simple example is $M(t) = \exp(-t/\tau_r)$ for $t \geq 0$, where τ_r is the decay-time of the material excitation that one wants to measure in the experiment. The coupling constant D depends on the material and the pump wavelength.

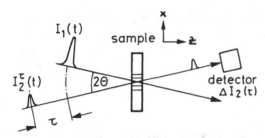

Fig. 7.14. Picosecond pump and probe experiment: the sample is excited by a powerful pump pulse with intensity $I_1(t)$ producing photo-induced absorption or bleaching. The sample transmission is monitored by a less intensive probe pulse $I_2^{\tau}(t)$ delayed by τ. The energy change of the transmitted probe pulse $\Delta I_2(\tau)$ is measured with a slow detector

$\Delta K(t)$ is probed with a pulse of intensity I_2^{τ} of the same shape as I_1 but delayed mechanically by τ: $I_2^{\tau}(t) = aI_1(t-\tau)$ where the intensity ratio $a \ll 1$. The change in the probe beam intensity in the sample with thickness d is

$$\Delta I_2^{\tau}(t) = aI_1(t-\tau) \cdot \{\exp(-Kd) - \exp[-(K+\Delta K)d]\}$$

$$\approx a\Delta K \, dI_1(t-\tau)\exp(-Kd) \ , \tag{7.30}$$

which is integrated over t by a slow detector and the measured signal is then

$$\Delta I_2(\tau) \approx a \, dD \int dt \, dt' M(t-t')I_1(t-\tau)I_1(t') \ . \tag{7.31}$$

To simplify the equations, it is assumed that $Kd, |\Delta K|d \ll 1$. Setting $t - t' = \tau' + \tau$, this equation can be rewritten as

$$\Delta I_2(\tau) \approx a \, dD \int d\tau' M(\tau' + t) \int dt' I_1(\tau' + t')I_1(t') \ , \tag{7.32}$$

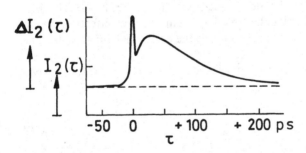

Fig. 7.15. Result of a picosecond sampling experiment according to Fig. 7.14 after *Shank* and *Auston* [7.73]: Energy $I_2(\tau)$ of the transmitted probe pulse vs. delay time τ. The material investigated was a Ge crystal . The interband absorption at 1.06 μm was bleached. The transmission decay is due to carrier recombination. The initial rise of the transmission is due to removal of electrons from the valence band by the time integrated intensity of the pump pulse. The sharp coherence peak at $\tau = 0$ is discussed in the text

which is just the impulse response to be measured convolved with the intensity autocorrelation function of the laser pulse.

However, this simple approach does not explain the coherence peak at $\tau = 0$ (Fig. 7.15). An explanation is obtained by considering the interaction of the pump and probe pulses which interfere because of their coherence, and produce a spatial modulation of both the absorption K and the refractive index, thus resulting in a transient grating. This diffracts the pump pulse into the direction of the probe beam. The phase of the diffracted pump beam is opposite to the phase of the probe, so that the probe beam amplitude is reduced. Diffraction at the transient grating therefore works like an additional photo-induced absorption, thus increasing the observed signal.

Interference of the pump and probe pulses takes place only if their delay τ is shorter than the coherence time t_c of the pulses, i.e. the intensity of the probe beam is increased by diffraction only in the regime $-t_c \leq \tau \leq t_c$, resulting in the coherence peak.

To calculate the amplitude and shape of the coherence peak, one starts with the time-independent wave equation for the total field strength E in the material

$$\frac{\partial^2 E}{\partial x^2} + \frac{\partial^2 E}{\partial z^2} + \varepsilon \frac{\omega^2}{c^2} E = 0 \ . \tag{7.33}$$

E is given by the pump $E_1(z, t)$ and probe $E_2(z, t)$ field strength, so that

$$E = [E_1(z, t)/2] \exp[\mathrm{i}(k_z z + k_x x)] + [E_2(z, t)/2] \exp[\mathrm{i}(-k_z z + k_x x)] \ . \tag{7.34}$$

The time dependence of $E_1(z, t)$ and $E_2(z, t)$ describes the temporal pulse shape and coherence of the pulses. The rapid time dependence due to the light frequency ω has been left out. The permittivity $\varepsilon = \varepsilon(x, z, t)$ is given by

$$\varepsilon = \varepsilon' + \mathrm{i}\varepsilon'' + \delta \int M(t - t')\overline{|E|^2} dt \ . \tag{7.35}$$

Here $(\varepsilon' + \mathrm{i}\varepsilon'')$ is the unperturbed complex permittivity, δ is a complex constant introduced to describe index and absorption changes due to the light energy density proportional to $\overline{|E|^2}$ with

$$\overline{|E|^2} = |E_1|^2 + |E_2|^2 + E_1 E_2^* \exp(\mathrm{i}2k_x x) + E_1^* E_2 \exp(-\mathrm{i}2k_x x) \ . \tag{7.36}$$

Because $\overline{|E|^2}$ is spatially modulated due to interference, ε also exhibits a grating-like structure. Introducing (7.34–36) into the wave equation (7.33) and assuming that the variation of $E_2(z, t)$ is negligible within a wavelength (i.e., neglecting the second derivative) yields for the change of the probe field strength

$$\mathrm{i}k_z \frac{\partial E_2}{\partial z} = -\mathrm{i}\varepsilon'' \frac{\omega^2}{c^2} E_2 - \delta \frac{\omega^2}{c^2} E_2 \int M |E_1|^2 dt' - \delta \frac{\omega^2}{c^2} E_1 \int M E_1^* E_2 dt' \ . \tag{7.37}$$

The first term on the right-hand side of this equation describes the unperturbed sample absorption, the second term gives an absorption increase due to the pump wave E_1 if δ is complex. The real part of δ gives a phase shift of the probe pulse, which is not considered here. Photo-induced absorption due to E_2 is neglected. The third term describes Bragg diffraction of the pump beam into the direction of the probe beam leading to the coherence peak.

From (7.37), the change in the probe-beam energy $\Delta I(\tau)$ due to photo-induced absorption, including the coherent coupling contribution, can be calculated by neglecting the variation of $E_1(z,t)$ and $E_2(z,t)$ within the sample, i.e. by putting $E_1(z,t) \approx E_1(0,t) = E_1(t)$ and $E_2(z,t) = \sqrt{a} E_1(t-\tau)$ on the right-hand side of (7.37). Starting from

$$\Delta I(\tau) = \int \Delta (1/2\,Z)|E_2|^2 dt = (1/2Z) \int (E_2^* \Delta E_2 + \text{c.c.}) dt \ , \tag{7.38}$$

$$\Delta E_2 \approx \left(\frac{\partial E_2}{\partial z}\right)_{\varepsilon''=c} d \ , \quad \text{one obtains} \tag{7.39}$$

$$\Delta I(\tau) = -(d/2Z)ka \, \text{Im}\left\{\delta(\gamma(\tau) + \beta(\tau))\right\} \ , \quad \text{with} \tag{7.40}$$

$$\gamma(\tau) = \iint E_1(t-\tau) E_1^*(t-\tau) M(t-t') E_1(t') E_1^*(t') dt \, dt' \ , \tag{7.41}$$

$$\beta(\tau) = \iint E_1(t) E_1^*(t-\tau) M(t-t') E_1^*(t') E_1(t'-\tau) dt \, dt' \ . \tag{7.42}$$

The term $\gamma(\tau)$ in (7.40) corresponds to the signal $\Delta I_2(\tau)$ given by (7.31) calculated by the simple approach. The second term $\beta(\tau)$ gives the coherent coupling contribution. This coherent coupling term depends only on $\text{Im}\{\delta\}$ which comes from diffraction at the amplitude grating formed by interference of the pump and probe beams. Diffraction at the phase grating, which comes from $\text{Re}\{\delta\}$ has not been considered. The reason is that diffraction at the phase grating produces a field strength contribution which is $90°$ out of phase with the probe field and which therefore gives only a second-order intensity change of the probe field.

Further evaluation of (7.42) is possible if the field strength of the pulse is written as a slowly varying envelope function $E(t)$ multiplied by a phase function or a random amplitude and phase function $f(t)$ to account for the coherence properties of the pulse (Sect. 7.6)

$$E_1(t) = E(t) f(t) \ . \tag{7.43}$$

Here $f(t)$ is characterized by the autocorrelation function $R(\tau)$

$$R(\tau) = \frac{1}{\Delta t} \int_{t'}^{t'+\Delta t} f(t) f^*(t-\tau) dt \ . \tag{7.44}$$

Assuming that $f(t)$ changes rapidly compared to $E(t)$, one obtains from (7.41, 42) by piecewise integration

$$\gamma(\tau) = |R(0)|^2 \int\int E^2(t-\tau) M(t-t') E(t) \, dt \, dt' \; , \tag{7.45}$$

$$\beta(\tau) = |R(\tau)|^2 \int\int E(t) E(t-\tau) M(t-t') E(t') E(t'-\tau) \, dt \, dt' \; . \tag{7.46}$$

In deriving (7.46), it has been assumed that the phase difference between E_1 and E_2 depends only on τ and not on t.

If $f(t)$ is changing rapidly compared to $E(t)$ (i.e., the coherence time is short compared to the pulse width), one obtains $\Gamma(\tau) \approx R(\tau)$ and $\beta(\tau) \approx |R(\tau)|^2 \approx |\Gamma(\tau)|^2$. The coherence peak measures then directly the absolute value of the coherence function of the pulse.

For further evaluation of the preceeding equations, it is necessary to make assumptions on the response function $A(t)$, and on the pulse field strength given by $E(t)$ and $f(t)$ or $R(\tau)$. A simple choice for $A(t)$ is the step function

$$M(t) = \begin{cases} 0 & t < 0 \; , \\ 1 & t \geq 0 \; . \end{cases} \tag{7.47}$$

This means that the absorption changes immediately with the incident intensity and that the decay of the absorption is neglected. An evaluation of (7.45, 46) with elementary functions is possible if the field strength is given by a one-sided exponential

$$E(t) = \begin{cases} 0 & t < 0 \; , \\ E_0 \exp(-\alpha t) & t \geq 0 \; . \end{cases} \tag{7.48}$$

The width of this pulse at half-maximum is given by

$$t_p = \ln 2 / 2\alpha \; . \tag{7.49}$$

A typical example for $R(\tau)$ is

$$R(\tau) = \exp(-v|\tau|) \tag{7.50}$$

where v is related to the coherence time by (Sect. 7.6):

$$t_c = \ln 2 / 2(\alpha + v) \; . \tag{7.51}$$

With (7.48, 50) one obtains

$$\gamma(\tau) = \begin{cases} \dfrac{E_0^4}{8\alpha^2} \exp 2\alpha\tau, & \tau < 0 \\[2mm] \dfrac{E_0^4}{8\alpha^2} [2 - \exp(-2\alpha\tau)], & \tau \geq 0 \; , \end{cases} \tag{7.52}$$

$$\beta(\tau) = \frac{E_0^4}{8\alpha^2} \exp[-2(\alpha + v)|\tau|] \; . \tag{7.53}$$

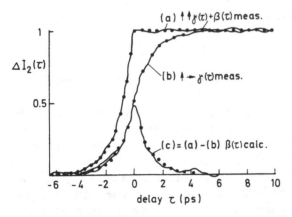

Fig. 7.16. Normalized change of the transmitted probe pulse energy $\Delta I_2(t)$ vs. delay time τ. The points give calculated values from (7.52, 53) for the case of maximum possible coherence ($v=0$). The lines have been obtained experimentally for (*a*) parallel polarizations of pump and probe pulses and (*b*) for perpendicular polarization. The maximum value of ΔI_2 corresponds to an absorption change of $\Delta K=8$ cm^{-1} for the thin amorphous silicon film used in the experiment of *Vardeny* and *Tauc* [7.75]

In Fig. 7.16, the calculated probe-pulse energy change given by $\gamma(\tau)+\beta(\tau)$ is shown for the case of maximum possible coherence of the pulse [$R(\tau)=$const, $v=0$]. The width of the pulse is taken to be $t_p=0.8$ ps. This pulse width has been chosen [7.75] to fit experimental results obtained with a passively mode-locked dye laser. The calculations agree very well with the experiments where photo-induced absorption in amorphous silicon has been investigated. Experimentally, it is possible to avoid coherent coupling and to measure $\gamma(\tau)$ alone by using a probe beam perpendicularly polarized with respect to the pump beam. Because the polarization memory of amorphous silicon is very short, the coupling between pump and probe beam is negligible for perpendicular polarization. The difference between the two measured signals gives the coherent coupling contribution $\beta(\tau)$ which agrees also with the simple model calculation.

It should be noted here that it is not always possible to avoid the coherence peak as has been done in the picosecond sampling experiment described for amorphous silicon. For example, in dye solutions, the polarization memory may be so long (see Sect. 3.4) that grating excitation is possible also with per-pendicularly polarized pump and probe beam. This means that a coherent coupling peak will be observed also in this case. The shape of the coherent coupling peak and its maximum value relative to the coherence-free signal has been calculated also for perpendicular polarizations and is found to be the same as calculated here [7.75].

Coherent coupling shall be discussed next for the case that the coherence time is smaller than expected from pulse width considerations [$v>0$ in (7.53)]. In this case, coherent coupling produces a distinct peak at $\tau=0$ as shown in Fig. 7.17.

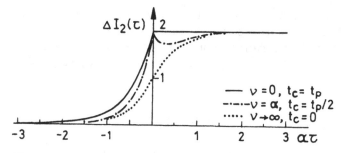

Fig. 7.17. Normalized change of the transmitted probe pulse energy $\Delta I_2(\tau)$ vs. delay time τ. The curves give calculated values from (7.52, 53) for different temporal coherence of the pulse

An experimental example is given in Fig. 7.15. Further examples are given in the literature [7.76–79]. Considerable differences of experimental coherence peaks compared to the peak shapes calculated here may occur due to other pulse shapes and coherence properties.

7.7.2 Coherence Peaks in Transient-Grating Experiments

The transient-grating method [7.80] for picosecond sampling experiments is sketched in Fig. 7.18. The two pump pulses 1 and 2 interfere and produce a transient grating in the sample, which is probed by pulse 3 at various delay times. An advantage of this arrangement compared to the photo-induced absorption technique is that the diffracted signal has no unwanted background. In absorption experiments, the small signal rides on the undisturbed probe pulse, so that signal averaging is necessary to discriminate the change of the probe pulse energy against the undisturbed background. This disadvantage can be overcome by background suppression with polarizers, thus exploiting the polarization rotation of the probe beam due to transient dichroism [7.76–79]. However, in transient dichroism experiments, one measures the absorption change squared [7.76] which is similar to transient grating experiments where the diffracted intensity is also given by the squared change of the absorption or refractive index.

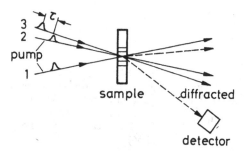

Fig. 7.18. Principle of picosecond transient grating apparatus

A coherent coupling effect in a transient–grating experiment comes from an additional grating which is produced by interference of beams 1 and 3 at zero delay time. Diffraction of beam 2 gives additional intensity in the direction of the detector, as can be proved easily [7.81]. The total diffracted energy reaching the detector is given by

$$J(\tau) = \left(\frac{\pi d}{2\lambda n}\right)^2 \frac{1}{2Z} \int_{-\infty}^{\infty} |E_3(t)\Delta\varepsilon_{12}(t) + E_2(t)\Delta\varepsilon_{13}(t)|^2 dt \ . \tag{7.54}$$

Here $\Delta\varepsilon_{12}$ is the amplitude of the (complex) permittivity modulation produced by pulses 1 and 2 and $\Delta\varepsilon_{13}$ is produced by pulses 1 and 3. Similar to (7.29), one obtains

$$\Delta\varepsilon_{12}(t) = \delta \int A(t-t')E_1^*(t')E_2(t')dt' \ ,$$
$$\Delta\varepsilon_{13}(t) = \delta \int A(t-t')E_1^*(t')E_3(t')dt' \ . \tag{7.55}$$

Further evaluation with one-sided exponential pulses according to (7.48), with an additional amplitude or phase modulation describing the coherence properties of the pulses according to (7.43, 44, 50) yields

$$J(\tau)/J(0) = \begin{cases} \exp(-2\alpha|\tau|)[1 + 3\exp(-2\nu|\tau|)], & \tau \leq 0 \\ 3 - 3\exp(-2\alpha\tau) + \exp(-4\alpha\tau) + 3\exp(-2\alpha\tau - 2\nu\tau), & \tau \geq 0 \ . \end{cases} \tag{7.56}$$

A numerical evaluation of this equation shows (Fig. 7.19) that a coherence peak appears already for $\nu = 0$, in contrast to Fig. 7.17. The coherence peak becomes sharper and more distinct with increasing ν, i.e. decreasing coherence of the pulse. Experimental examples of coherence peaks in picosecond transient grating experiments have been given in [7.81].

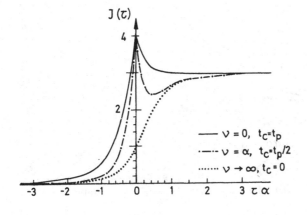

—— $\nu = 0$, $t_c = t_p$
—··— $\nu = \alpha$, $t_c = t_p/2$
······ $\nu \to \infty$, $t_c = 0$

Fig. 7.19. Normalized energy of the diffracted probe pulse in a transient grating experiment according to Fig. 7.18, calculated from (7.56)

7.7.3 Additional Contributions to the Coherent Coupling Peak

The peak appearing at zero delay time in picosecond sampling experiments may be caused also by mechanisms other than grating excitation.

An additional excitation of the material with an optical response faster than the pulse width will contribute to the peak amplitude. Such a situation has been investigated [7.79] in dye solutions where the 5 ps pump pulse first excites higher vibrational S_1 states, which relax in less than 1 ps to the lowest excited singlet state S_1. The following relaxation is much slower. The complex index change due to the initial excitation is larger than the change produced by the population in the lowest S_1 state because the initiallly populated states are in resonance with the frequency of the probing pulse.

A second mechanism is nonresonant four-wave mixing. In dye solutions, this may take place in the solvent and give a contribution to the zero delay time peak. A similar situation may occur also in other materials like semiconductors, if one wants to make time resolved measurements on the population of a state which produces only a weak complex index change. A large nonresonant third-order susceptibility due to virtual transitions to other states may result in relatively strong four-wave mixing and contribute to the zero delay time peak.

In summary, the coherence peak or better zero delay-time peak may be composed of several contributions, like an unresolved spectral line, and care has to be taken to extract physical information from this peak.

7.8 Doppler-Free Spectroscopy by DFWM

Population gratings and degenerate four-wave mixing (DFWM) have been investigated not only in liquids and solids, as outlined in Sect. 3.4, but also in gases and vapors. Experiments have been performed, e.g., in Na and Hg vapors with cw dye lasers [7.82–89] and in SF_6 and CO_2 gas with CO_2 lasers [7.90–94]. For a complete description of these gratings in gases, it is necesssary to use quantum theory which includes also coherence effects similarly to Sect. 3.3.

At a first glance, one may expect that population gratings in gases decay rapidly due to the thermal motion of the atoms or molecules. However, this is not true because only atoms moving in the grating planes interact efficiently with the light fields. This can be understood as follows.

The frequency ω of a light wave incident on a moving atom with velocity v is shifted due to the Doppler effect, so that the atom sees a frequency $\omega' = \omega - k \cdot v$, k being the light wave vector. In a four-wave mixing arrangement (Sect. 4.6), three waves with equal frequencies and wave vectors k_1, $k_2 = -k_1$ and k_4 are incident. Due to the Doppler effect, an atom generally sees three different frequencies. However, interaction of the atom with the light is most efficient if these frequencies are equal to the resonance frequency ω_0 of the atom

$$\omega_0 = \omega - k_1 \cdot v \ ,$$

$$\omega_0 = \omega - k_2 \cdot v \ ,$$

$$\omega_0 = \omega - k_4 \cdot v \ .$$

These equations imply that $\omega = \omega_0$ and that the velocity of the atoms has to be perpendicular to the plane given by k_1 and k_4. Then the Doppler shift for all three waves is zero. This means that the grating is built up mainly by atoms moving parallel to the grating planes. If collisions are neglected, such a grating decays only due to the lifetime of the atoms and not due to thermal motion.

The velocity selection in the grating build-up has an interesting spectroscopic application. Therefore the grating is excited and detected with a tunable laser frequency ω. The grating amplitude and the diffracted wave with wave-vector $k_3 = -k_4$ peaks at $\omega = \omega_0$ and decreases at smaller and larger frequencies. In the simplest case, when thermal motion is small, the spectrum of the signal wave has a width corresponding to the natural line width (inverse lifetime) of the atoms. Doppler-broadening is eliminated.

The detailed line-shapes of the DFWM spectra can be quite complicated due to saturation effects [7.84, 85, 94]. Also thermal motion has to be considered [7.95].

More detailed discussions of Doppler-free laser spectroscopy via degenerate four-wave mixing have been given in [7.96, 97].

8. Conclusion and Outlook

The consideration of laser-induced dynamic gratings allows a pictorial description of various, in particular third-order nonlinear optical effects. Gratings are important in different fields of optical science like holography, light scattering, acousto-optics and spectroscopy. The combination of these techniques with dynamic gratings and nonlinear wave-mixing effects has widened the possibilities of optical measuring methods and lead to the demonstration of new devices. We expect further future development in the directions outlined in the following.

8.1 Light-Deflection and Modulation

The most direct application of laser-induced gratings is the use in light-deflection and modulation devices in analogy to acousto-optic systems. With solid-state or gas lasers as excitation sources, such devices are bulky and expensive. However, the use of semiconductor lasers could result in a considerable simplification. A semiconductor laser may be better comparable in size and cost to an acousto-optic transducer.

Compared to acousto-optic light deflection, laser induced gratings allow much shorter response times. The main attraction of laser-induced grating deflectors may be in the field of integrated optics [8.1, 2], and all-optical systems for information processing. Components exploiting laser-induced gratings can be fabricated from the same materials which are used for waveguides, lasers, modulators, bistable and other devices. To evaluate these possibilities, experimental and theoretical work on laser-induced grating phenomena in optical waveguides is highly desirable.

8.2 Information Processing, Real-Time Holography and Phase Conjugation

Applications of coherent optical methods in information processing [8.3, 4] have to compete with the presently available and rapidly developing electronic computers. It seems unlikely that this mature technology could be replaced by optical computers in the near future. However, optical systems may find a technological niche for special applications requiring a large number of special mathematical operations like Fourier transforms, correlations and multiplication of matrices [8.5]. Direct optical real-time processing or preprocessing of two-dimensional arrays of information (pictures) [8.6], e.g. in optical object recognition systems, could lead to a significant data reduction for further electronic processing. Thus hybrid opto-electronic systems seem to be of increasing interest. Similarly the future of real-time holography and phase conjugation depends on the incorporation of these ideas into routinely working systems, as well as on the development of suitable nonlinear optical materials of high quality and sufficient sizes.

8.3 Forced Light Scattering

Like other optical methods, forced light scattering has the advantage that measurements can be done without physical contact and on a microscopic scale. High time resolution is obtained by using pico- and femtosecond laser pulses for grating excitation [8.7]. With lasers becoming conveniently available measuring tools, advanced optical test methods should develop into standard laboratory equipment, like electronic or microwave test apparatus today. Further development of laser applications in material-processing and optical-communication system will enhance the need of optical test apparatus and foster the utilization of such systems. In this context, forced light scattering is becoming an established technique to measure material properties, e.g., diffusion constants in various systems and nonlinear optical refraction and absorption. Examples have been given extensively in this monograph.

List of Symbols

Some symbols with only local importance are omitted

Roman Letters

a	Intensity ratio of pump and probe beams (Sect. 7.6)
A	Electric field amplitude of a light wave (see text for meaning of indices)
c	Velocity of light (in vacuum)
c	Specific heat
c.c.	Complex conjugate of the preceding expression
c_p	Specific heat at constant pressure (Sect. 3.7)
C	Concentration
C	Coupling constant (Sect. 3.4)
d	Thickness
D	Coupling constant (Sect. 7.6)
D	Electron diffusion constant (Sect. 3.6)
D	Rotational diffusion constant (Sect. 3.4)
D	Energy diffusion constant (Sect. 5.6)
D_a	Ambipolar diffusion constant
D_i	Electric displacement (Sect. 5.3)
D_m	Mass diffusivity
D_{th}	Thermal diffusivity
e	Elementary charge
E	Electric field strength (see text for meaning of indices)
E_b	Exciton binding energy
E_g	Energy gap of a semiconductor
E_p	Light energy area density ($E_p = \int I\,dt$)
E_0, E_1, E_2	Energies of electronic states
f	Sound frequency (Sect. 5.3)
F	Total power or light flux of a laser beam
f_a, f_b	Population probability
$f(t)$	Amplitude or phase modulation of a pulse (Sect. 7.5)
f_{ijkl}	Coupling coefficient (Sect. 2.3)
g	Generation rate
g^p	Coupling coefficient (Sect. 2.3)
h	Peak to groove height of a surface grating (Sect. 4)
h	Planck's constant $= 6.626 \times 10^{-34}\,Js$
$\hbar$	Planck's constant $h/2\pi$

i	$\sqrt{-1}$
I	Intensity (light power area density), (see text for meaning of indices)
I_0	Peak intensity of a (Gaussian) beam in space (Sect. 2)
$\hat{I}$	Peak intensity of a laser pulse in time (Sect. 2)
J	Current density (Sect. 3.6)
J	Energy of a diffracted pulse (Sect. 7.5)
J_m	mth order Bessel function
k	Wave vector (see text for meaning of indices)
k_B	Boltzmann constant
k_T	Soret constant (Sect. 3.8)
K	Absorption coefficient $[I/I_0 = \exp(-Kd)]$
L	Length
m	Modulation ratio (Sect. 3.6)
m	Diffraction order number
m	Free electron mass (Sect. 3.5)
m_e, m_h	Electron, hole mass
m_{eh}	Reduced effective mass $1/m_{eh} = 1/m_e + 1/m_h$
M_1, M_2	Mirrors (Sect. 7.1)
$M(t)$	Impulse response of a medium (Sect. 7.6)
n	Refractive index $n = \mathrm{Re}\{\tilde{n}\}$
$\tilde{n}$	Complex refractive index
n_e	Electron density (Sect. 3.6)
n_{eh}	Refractive index change per electron-hole pair
n_2	Nonlinear refractive index $n_2 = \mathrm{Re}\{\tilde{n}_2\}$
n_3	Nonlinear refractive index related to field strength (Sect. 2.7)
N	Particle density (see text for meaning of different indices)
N	Electron-hole density (Sect. 5.5)
N_A	Acceptor (trap) density
N_D	Donor density
p	Longitudinal laser mode number (Sect. 7.1)
p	Polarization (dipole density), (see text for meaning of indices)
P_t	Total power of a light beam (see text for meaning of other indices)
P_{diff}	Diffracted power
p_{ijkl}	Photoelastic constants (Sect. 5.3)
q	Grating vector
Q	Grating thickness parameter
r_{ij}	Electro-optic tensor
r	Position vector, radius vector
$R(z)$	Electric field amplitude of a light wave in a self-diffraction experiment (Sect. 4.5)
$R(\tau)$	Autocorrelation function of $f(t)$
s_{ij}	Velocity gradient tensor (Sect. 5.4)
S	Amplitude of a partial wave
$S(z)$	Electric field amplitude of a light wave in a self-diffraction experiment (Sect. 4.5)

t	Time
t_c	Coherence time (Sect. 7.5)
t_p	Pulse width
$t(x)$	Amplitude transmittance (Sect. 4)
T	Intensity transmittance (Sect. 4)
T	Temperature
T_1	Excited state lifetime
T_2	Polarization decay time, phase relaxation time
u_i	Material displacements (Sect. 5.3)
u_{ik}	Strain
v	Sound velocity (Sect. 5.3)
v	Flow velocity (Sect. 5.4)
V	Volume (Sect. 3.4)
V	Voltage (Sect. 3.6)
w	Spot size radius of a TEM_{00} laser beam (Sect. 2)
W	Total laser pulse energy
x	Fraction of components in an alloy (Sect. 3.5)
x	Coordinate in grating direction
x_1, x_2, x_3	Cartesian coordinates (Sect. 5.4)
X	Grating excitation (e.g. temperature, density)
z	Coordinate perpendicular to grating
z_0	Overlap length to two light beams
Z	Wave impedance

Greek Letters

α_0, α_1	atomic (molecular) polarizability
α_{ik}	Particle polarizability (Sect. 3.7)
α_l	Linear thermal expansion coefficient (Sects. 3.7, 5.2)
α	Angle of incidence
α	Pulse width parameter (Sects. 7.5, 7.6)
α, β	Fit parameters (Sect. 5)
β	Cubic thermal expansion coefficient (Sect. 5.3)
β	Thermal excitation rate (Sect. 3.6)
$\beta(\tau)$	Coherent coupling contribution to coherence peak (Sect. 7.6)
$\gamma(\tau)$	Signal observed in a transient absorption sampling experiment without coherent coupling (Sect. 7.6)
γ	Population density difference
γ	Ratio of specific heats c_p/c_v (Sect. 5.3)
γ_R	Recombination constant
Γ	Coherence function (Sect. 7.5)
Γ_R	Second sound damping constant (Sect. 3.7)
δ	Coupling constant (Sect. 7.6)
$\delta n_{A,B}$	Refractive index per mole of molecules A or B in a solution (Sect. 5.8)

$\varDelta$	Small change of a quantity
$\varDelta M$	Interference tensor
ε	Relative (electric) permittivity
ε_1	Spatial amplitude of relative permittivity
ε_0	Absolute permittivity of vacuum
ζ	Quantum efficiency
η	Diffraction efficiency
η	Viscosity (Sect. 5.3)
θ	Intersection angle of two light beams
θ	Polar coordinate, angle giving orientation of a molecule (Sect. 3.4)
κ	Photovoltaic constant (Sect. 3.6)
κ	Coupling constant (different definitions in Sects. 4.4, 4.5, 4.6)
λ	Vacuum wavelength
λ	Thermal conductivity (Sect. 3.7, 5.3)
λ_{ijkl}	Elastic stiffness constants (Sect. 5.3)
$\varLambda$	Grating period
μ	(Magnetic) permeability
μ	Transition dipole moment (Sect. 3.4)
μ	Mobility (Sect. 3.6)
ν	Frequency
ν	Number of phase jumps per unit time (Sect. 7.5)
ϱ	Density
ϱ	Cylindrical coordinate perpendicular to propagation direction of a laser beam (Sect. 2)
ϱ	Orientational distribution of molecules (Sect. 3.4)
ϱ	Electromagnetic energy density (Sect. 7.1)
σ	Absorption cross section
σ_m	Wave-vector of a partial wave
σ_{ik}	Stress
τ	Delay time
τ	Recombination time (Sects. 3.5, 3.6)
τ_d	Damping time
τ_D	Diffusive decay time
τ_f	Fluorescence decay time (Sect. 5.6)
τ_g	Grating decay time (Sect. 5.5, 5.6)
τ_m	Relaxation time of a concentration grating (Sect. 3.8)
τ_{or}	Orientational decay time
τ_q	Relaxation time of a temperature grating (Sects. 3.7, 3.8)
τ_r	Relaxation time of a material excitation (Sect. 7.6)
τ_s	Singlet lifetime
τ_w	Relaxation time of a temperature distribution with width w (Sect. 3.7)
τ_0	Delay time (Sect. 7.5)
τ_0	Average amplitude transmittance (Sect. 4)
τ_1	Spatial amplitude of amplitude transmittance (Sect. 4)

τ_1, τ_2	Lifetimes of energy levels (Sect. 5)
φ	Phase angle describing grating shift (Sect. 4)
ϕ	Polar coordinate, angle giving orientation of a molecule (Sect. 3.4)
ϕ	Phase of a light wave (Sect. 7.5)
ϕ	Phase shift (Sects. 2, 4)
ϕ	Grating slant angle (Sects. 3.6, 4)
ϕ_m	Diffraction angle
χ	Susceptibility
$\chi^{(3)}$	Third-order nonlinear susceptibility
ψ	Phase shift between reference and diffracted wave (Sect. 2.4.5)
ω	Circular frequency (see text for meaning of indices)
Ω	Rabi frequency (Sect. 3.3)
Ω	Modulation frequency (Sect. 3.7)
Ω	Difference frequency (Sect. 4.7)
Ω_0	Second sound frequency (Sect. 3.7)

Units

SI units (MKSA units) are used if not stated otherwise
Temperature is given in K

Relation between light intensity and electric field strength amplitude

$$I = E^2/2Z \quad \text{with} \quad Z = \text{wave impedance} = \sqrt{\mu\mu_0/\varepsilon\varepsilon_0}$$

Mathematical Symbols

Vectors are in boldface type (e.g. $\boldsymbol{E}$ = electric field strength, E = magnitude of $\boldsymbol{E}$)

$\parallel$	Parallel
$\perp$	Perpendicular
$\lvert\chi\rvert$	Absolute value of a complex quantity χ
$\text{Re}\{\chi\}$, $\text{Im}\{\chi\}$	Real, imaginary part of complex quantity χ
χ^*	Complex conjugate of χ

References

Chapter 1

1.1   O. Wiener: Ann. Phys. (Leipzig) **40**, 203 (1890);
M. Born, E. Wolf: *Principles of Optics* (Pergamon, Oxford 1975) p. 279

1.2   G. Lippmann: Compt. Rend. Seanc. Acad. Sci. Paris **112**, 274 (1891); **114**, 124 (1892)

1.3   D. Gabor: Nature (London) **161**, 77 (1948);
E. N. Leith, J. Upatnieks: J. Opt. Soc. Am. **53**, 1377 (1963);
R. J. Collier, C. B. Burkhardt, L. H. Lin: *Optical Holography* (Academic, New York, 1971);
H. J. Eichler: *Bergman-Schaefer, Lehrbuch der Exp.-phys. Band III*, Optik (de Gruyter, Berlin 1974) p. 422;
H. J. Caulfield: *Handbook of Optical Holography* (Academic, New York 1979)

1.4   H. M. Smith: *Holographic Recording Materials*, Topics Appl. Phys., Vol. 20 (Springer, Berlin, Heidelberg 1977)

1.5   B. Ya. Zel'dovich, N. F. Philipetsky, V. V. Shkunov: *Principles of Phase Conjugation*, Springer Ser. Opt. Sci., Vol. 42 (Springer, Berlin, Heidelberg 1985)

1.6   P. Debye, F. W. Sears: Proc. Nat. Acad. Sci. **18**, 409 (1932);
R. Lucas, P. Biquard: J. Phys. Rad. **3**, 464 (1932)

1.7   H. Haken, H. Sauermann: Z. Phys. **173**, 261 (1963);
See H. Haken: *Laser Theory* (Springer, Berlin, Heidelberg 1984)

1.8   C. L. Tang, H. Statz, G. de Mars: J. Appl. Phys. **34**, 2289 (1963)

1.9   J. M. Green, J. P. Hohimer, F. K. Tittel: Opt. Commun. **7**, 349 (1973);
H. W. Schroeder, L. Stein, D. Froehlich, B. Fugger, H. Welling: Appl. Phys. **14**, 377 (1977);
See also W. Demtroeder: *Laser Spectroscopy*, Springer Ser. Chem. Phys., Vol. 5 (Springer, Berlin, Heidelberg 1981)

1.10 H. Boersch, H. J. Eichler: Z. Angew. Phys. **22**, 378 (1967)

1.11 N. Bloembergen: *Nonlinear Optics* (Benjamin, New York 1977);
S. A. Akhmanov, N. I. Koroteev: "Nonlinear Optical Techniques in Spectroscopy of Light Scattering", in Series *Problems in Modern Physics* (Nauka, Moscow 1981) (in Russian);
Y. R. Shen: *The Principles of Nonlinear Optics* (Wiley, New York 1984)

1.12 W. Kaiser, M. Maier: "Stimulated Rayleigh, Brillouin and Raman Spectroscopy", in *Laser Handbook*, Vol. 2, ed. by F. T. Arechi, E. O. Schulz-Dubois (North-Holland, Amsterdam 1972)

1.13 I. P. Batra, R. H. Enns, D. W. Pohl: Phys. Status Solidi (b) **48**, 11 (1971)

1.14 R. Figgins: Contemp. Phys. **12**, 283 (1971)

1.15 Y. R. Shen: Rev. Mod. Phys. **48**, 1 (1976)

1.16 N. Bloembergen, G. Bret, P. Lallemand, A. S. Pine, P. Sinowa: IEEE J. QE, **3**, 197 (1967)

1.17 M. Denariez, G. Bret: Phys. Rev. **171**, 1601 (1968)

1.18 D. W. Pohl, M. Maier, W. Kaiser: Phys. Rev. Lett. **20**, 366 (1968);
D. W. Pohl, I. Reinhold, W. Kaiser: Phys. Rev. Lett. **20**, 1461 (1968);
W. Rother, D. W. Pohl, W. Kaiser: Phys. Rev. Lett. **22**, 915 (1969);
D. W. Pohl: Phys. Rev. Lett. **23**, 711 (1969)

1.19 H. J. Eichler, G. Salje, H. Stahl: J. Appl. Phys. **44**, 5383 (1973);
      See also H. J. Eichler, G. Enterlein, P. Glozbach, J. Munschau, H. Stahl: Appl. Opt. **11**, 372 (1972)
1.20 D. W. Pohl, S. E. Schwarz, V. Irniger: Phys. Rev. Lett. **31**, 32 (1973)
1.21 J. P. Woerdman, B. Boelger: Phys. Lett. **30A**, 164 (1969);
      J. P. Woerdman: Philips Res. Rep. Suppl. No. 7 (1971)
1.22 D. W. Phillion, D. J. Kuizenga, A. E. Siegmann: Appl. Phys. Lett. **27**, 85 (1975)
1.23 R. Y. Chiao, P. L. Kelley, E. Garmire: Phys. Rev. Lett. **17**, 1158 (1966);
      R. L. Carman, R. Y. Chiao, P. L. Kelley: Phys. Rev. Lett. **17**, 1281 (1966)
1.24 H. J. Gerritsen, E. G. Ramberg, S. Freeman: *Proc. Symp. Mod. Opt.*, ed. by J. Fox
      (Polytechnic Press, New York 1967) p. 109
1.25 B. I. Stepanov, E. V. Ivakin, A. S. Rubanov: Sov. Phys.-Dokl. **16**, 46 (1971)
1.26 B. Ya. Zel'dovich, V. I. Popovichev, V. V. Ragul'skii, F. S. Farsullov: JETP Lett. **16**, 435 (1972)
1.27 A. Yariv: Appl. Phys. Lett. **28**, 88 (1967);
      See also A. Yariv: J. Opt. Soc. Am. **66**, 301 (1976)
1.28 R. W. Hellwarth: J. Opt. Soc. Am. **67**, 1 (1977)
1.29 J. Feinberg: Opt. Lett. **5**, 330 (1980)
1.30 P. Günter: Opt. Lett. **7**, 10 (1982)
1.31 R. A. Fisher (ed.): *Optical Phase Conjugation* (Academic, New York 1983)
1.32 D. M. Pepper (ed.): Special Issue on Nonlinear Optical Phase Conjugation, Opt. Eng. **21**, 2 (1982)
1.33 P. Günter: Phys. Rept. **93**, 199–299 (1982)

Chapter 2

2.1   See, for instance, F. T. Arechi, E. O. Schulz-Dubois (eds.): *Laser Handbook*, Vol. 1
      (North-Holland, Amsterdam 1972);
      M. Sargent III, M. O. Scully, W. E. Lamb Jr.: *Laser Physics* (Addison-Wesley, Reading, MA 1974);
      O. Svelto: *Principles of Lasers*, 2nd ed. (Plenum, New York 1982);
      H. Haken: *Laser Theory* (Springer, Berlin, Heidelberg 1984)
2.2   D. B. Brayton: Appl. Opt. **13**, 2346 (1974)
2.3   A. E. Siegman: J. Opt. Soc. Am. **67**, 545 (1977)
2.4   C. V. Shank, R. L. Fork, F. Beisser: Laser Focus **19**, 59 (1983)
2.5   W. Born, E. Wolf: *Principles of Optics*, 6th ed. (Pergamon, London 1980)
2.6   J. P. Woerdman: Philips Res. Rep. Suppl. No. 7 (1971);
      See also J. P. Woerdman: Opt. Commun. **2**, 212 (1970)
2.7   See, for instance, R. M. Gagliardi, S. Karp: *Optical Communications* (Wiley, New York 1976)
2.8   D. W. Pohl: IBM J. Res. Devel. **23**, 604 (1979)
2.9   For an introduction to laser velocimetry, see, for example, T. S. Durrani, C. A.
      Greated: *Laser systems in flow measurements*, (Plenum, New York 1977);
      F. Durst, A. Melling, J. H. Whitelaw: *Principles and Practice of Laser-Doppler-anemometry* (Academic, New York 1975)
2.10 See, for example, E. O. Schulz-Du Bois (ed.): *Photon Correlation Techniques*, Springer
      Ser. Opt. Sci., Vol. 38 (Springer, Berlin, Heidelberg 1983)
2.11 D. W. Phillion, D. J. Kuizenga, A. E. Siegman: Appl. Phys. Lett. **27**, 85 (1975)
2.12 D. W. Pohl: Phys. Lett. **77A**, 53 (1980)
2.13 R. L. Carman, R. Y. Chiao, P. L. Kelley: Phys. Rev. Lett. **17**, 1281 (1966);
      F. Gires: CR Acad. Sci., Ser. B, **266**, 596 (1968);
      M. E. Mack: Appl. Phys. Lett. **12**, 329 (1968) and Phys. Rev. Lett. **22**, 13 (1969)
2.14 K. O. Hill: Appl. Opt. **10**, 1695 (1971)

2.15   D. W. Pohl, I. Reinhold, W. Kaiser: Phys. Rev. Lett. **20**, 1141 (1968);
        W. Rother, D. W. Pohl, W. Kaiser: Phys. Rev. Lett. **22**, 915 (1969);
        W. Rother, H. Meyer, W. Kaiser: Z. Naturforsch. A, **25**, 1136 (1970);
        W. Rother: Z. Naturforsch. A, **25**, 1120 (1970);
        W. Rother, W. Meyer, W. Kaiser: Phys. Lett. **31A**, 245 (1970)
2.16   S. A. Akhmanov, N. I. Koroteev: Sov. Phys. Usp. **20**, 899 (1978);
        A. Krumins, P. Günter: Appl. Phys. **19**, 153 (1979);
        P. Günter: Phys. Rept. **93**, 199–299 (1982)
2.17   H. Fery: Ph. D. Thesis, Technische Universitaet Berlin, (1975) D 83;
        See also H. J. Eichler: Opt. Acta **24**, 631 (1977);
        Z. Vardeny, J. Tauc: Opt. Commun. **39**, 396 (1981)
2.18   C. Allain, H. Z. Cummins, P. Lallemand: J. Physique **39**, L-473 (1978)
2.19   D. W. Pohl, S. Schwarz: Phys. Rev. Lett. **31**, 32 (1973)
2.20   J. A. Armstrong, N. Bloembergen, J. Ducuing, P. S. Pershan: Phys. Rev. **127**, 1918
        (1962)

Chapter 3

3.1   E. I. Shtyrkov: Opt. Spektr. **45**, 603 (1978)
3.2   E. I. Shtyrkov, V. S. Lobkov, N. G. Yarmukhametov: JETP Lett. **27**, 648 (1978)
3.3   E. I. Shtyrkov, N. L. Nevelskaya, V. S. Lobkov, N. G. Yarmukhametov: Phys. Status
        Solidi (b) **98**, 473 (1980);
        Further related work: E. I. Shtyrkov, V. S. Lobkov, S. A. Moiseev, N. G.
        Yarmukhametov: Sov. Phys. JETP **54**, 1041 (1981)
3.4   R. H. Pantell, H. E. Puthoff: *Fundamentals of Quantum Electronics* (Wiley, New York
        1969) p. 56
3.5   T. W. Mossberg, R. Kachru, S. R. Hartmann, A. M. Flusberg: Phys. Rev. A**20**, 1976
        (1979)
3.6   E. Courtens: In *Laser Handbook*, Vol. 2 (North-Holland, Amsterdam 1972) p. 1259
3.7   T. W. Mossberg, R. Kachru, E. Whittaker, S. R. Hartmann: Phys. Rev. Lett. **43**, 851
        (1979)
3.8   C. V. Heer, R. L. Sutherland: Phys. Rev. A**19**, 2026 (1979)
3.9   W. H. Hesselink, D. A. Wiersma: Phys. Rev. Lett. **43**, 1991 (1979)
3.10   C. V. Heer, N. C. Griffin: Opt. Lett. **4**, 241 (1979)
3.11   T. Yajmia, Y. Taira: J. Phys. Soc. Jap. **47**, 1620 (1979)
3.12   N. V. Kukhtarev, T. I. Semenets: Qantovaja Elektronika **7**, 1721 (1980)
3.13   P. Aubourg, J. P. Bettini, G. P. Agarwal, P. Cottin, D. Guerin, O. Meunier, J. L.
        Boulnois: Opt. Lett. **6**, 383 (1981)
3.14   J. F. Lam, D. G. Steel, R. A. Mc Farlane, R. C. Lind: Appl. Phys. Lett. **38**, 977 (1981)
3.15   J. P. Woerdman, M. F. H. Schuurmans: Opt. Lett. **6**, 239 (1981)
3.16   R. W. Ditchburn: *Light* (Academic, London 1976) p. 735;
        C. Kunz (ed.): *Synchrotron Radiation*, Topics Current Phys., Vol. 10 (Springer, Berlin,
        Heidelberg 1979) p. 172
3.17   M. Schubert, B. Wilhelmi: *Einführung in die Nichtlineare Optik*, Teil II (B. G. Teubner
        Verlagsgesellschaft, Leipzig 1971)
3.18   H. Boersch, H. J. Eichler: Z. Angew. Phys. **22**, 378 (1967)
3.19   T. H. Maiman, R. H. Hoskins, I. T. D'Haenens, C. K. Asawa, V. Evtuhov: Phys. Rev.
        123, 1151 (1961)
3.20   O. C. Cronemeyer: J. Opt. Soc. Am. **56**, 1703 (1964)
3.21   W. Koechner: *Solid-State Laser Engineering*, Springer Ser. Opt. Sci., Vol. 1 (Springer,
        New York 1976)
3.22   K. C. Kiang, J. G. Stephany, F. C. Unterleitner: IEEE J. QE-1, 295 (1966)
3.23   T. Kushida: J. Phys. Soc. Jap. **21**, 1331 (1966)

234   References

3.24   A. M. Bonch-Bruevich, T. K. Razumova, Ya. A. Imas: Opt. Spectr. **20**, 575 (1966)
3.25   A. Szabo: Opt. Commun. **12**, 366 (1974)
3.26   W. M. Fairbank, G. K. Klauminzer, A. L. Schawlow: Phys. Rev. **B11**, 60 (1975)
3.27   I. S. Gorban, G. L. Kononchuk: Opt. Spect. **17**, 478 (1964)
3.28   N. K. Belskii, D. A. Mukhamedova: Sov. Phys.-Doklady **9**, 798 (1965)
3.29   H. Weber: „Das Emissionsverhalten des gepulsten Rubinlasers", Habilitationsschrift, Technische Universität Berlin (1967)
3.30   D. Pohl: Phys. Lett. **26A**, 357 (1968)
3.31   P. F. Liao, D. M. Bloom: Opt. Lett. **3**, 5 (1978)
3.32   H. J. Eichler, P. Glozbach, B. Kluzowski: Z. Angew. Phys. **28**, 303 (1970)
3.33   H. J. Eichler, J. Eichler, J. Knof, C. H. Noack: Phys. Status Solidi **52**, 481 (1979)
3.34   D. S. Hamilton, D. Heiman, I. Feinberg, R. W. Hellwarth: Opt. Lett. **4**, 124 (1979)
3.35   P. F. Liao, L. M. Humphrey, D. M. Bloom, S. Geschwind: Phys. Rev. **B20**, 4145 (1979)
3.36   P. E. Jessop, A. Szabo: Phys. Rev. Lett. **45**, 1712 (1980)
3.37   S. Chu, H. M. Gibbs, S. L. Mc Call, A. Passner: Phys. Rev. Lett. **45**, 1715 (1980)
3.38   K. O. Hill: Appl. Opt. **10**, 1695 (1959)
3.39   J. R. Salcedo, A. E. Siegmann, D. D. Dlott, M. D. Fayer: Phys. Rev. Lett. **41**, 131 (1978)
3.40   Ch. M. Lawson, R. C. Powell, W. Zwicker: Phys. Rev. Lett. **46**, 1020 (1981)
3.41   A. Tomita: Appl. Phys. Lett. **34**, 463 (1979)
3.42   H. Park, W. H. Steier: IEEE J. **QE-17**, 581 (1981)
3.43   L. F. Mollenauer: "Color Center Lasers", in Methods of Appl. Phys., Vol. 15, Quantum Electronics, ed. by C. L. Tang (Academic, New York 1979)
3.44   R. C. Duncan, D. L. Staebler: "Inorganic photochromic materials", in Holographic Recording Materials, ed. by H. M. Smith, Topics Appl. Phys., Vol. 20 (Springer, Berlin, Heidelberg 1977)
3.45   J. M. Wiesenfeld, L. F. Mollenauer, E. I. Ippen: Phys. Rev. Lett. **47**, 1668 (1981)
3.46   H. J. Gerritsen, E. G. Ramberg, S. Freeman: "Image Processing with nonlinear optics", in *Proc. Symp. Modern Optics*, ed. by J. Fox (Polytechnic, New York 1967)
3.47   H. J. Eichler, B. Kluzowski: Z. Angew. Phys. **27**, 4 (1969)
3.48   M. E. Mack: Phys. Rev. Lett. **22**, 13 (1969);
       See also Appl. Phys. Lett. **12**, 329 (1968)
3.49   E. I. Shtyrkov: ZhETF Pis. Red. **12**, 92 (1970)
3.50   H. J. Eichler, G. Enterlein, J. Munschau, H. Stahl: Z. Angew. Phys. **31**, 1 (1971)
3.51   R. I. Scarlet: Phys. Rev. **A6**, 2281 (1972)
3.52   H. J. Eichler, G. Enterlein, P. Glotzbach, J. Munschau, H. Stahl: Appl. Opt. **11**, 372 (1972)
3.53   H. E. Lessing, A. von Jena: "Continuous picosecond spectroscopy of dyes", in *Laser Handbook*, Vol. 3, ed. by M. L. Stitch (North-Holland, Amsterdam 1979)
3.54   D. W. Phillion, D. J. Kuizenga, A. E. Siegman: Appl. Phys. Lett. **27**, 85 (1975)
3.55   D. Langhans: „Untersuchungen an transienten laserinduzierten Gittern im Pikose-kundenbereich", Dissertation, Universität Berlin (1980)
3.56   A. von Jena, H. E. Lessing: Opt. and Quant. Electr. **11**, 419 (1979)
3.57   A. von Jena: Appl. Phys. **B26**, 1 (1981)
3.58   Y. Silberberg, I. Bar-Joseph: IEEE J. **QE-17**, 1967 (1981)
3.59   T. Todorov, L. Nikolova, N. Tomova, V. Dragostinova: Opt. and Quant. Electr. **13**, 209 (1981)
3.60   D. Magde: IEEE J. **QE-17**, 489 (1981)
3.61   N. Wiese: „Absorptions- und Brechungsindexänderungen bei der Erzeugung lichtinduzierter Gitter in Farbstofflösungen", Diplomarbeit, Technische Universität Berlin (1984), to be published
3.62   R. L. Fork, B. I. Greene, C. V. Shank: Appl. Phys. Lett. **38**, 671 (1981)
3.63   J. Vaitkus, K. Jarasiunas: Sov. Phys. Coll. **19**, 32 (1979)
3.64   A. Miller, D. A. B. Miller, S. D. Smith: Adv. Phys. **30**, 697 (1981)
3.65   R. K. Jain: Opt. Eng. **21**, 199 (1982)

3.66   J. P. Woerdman: Philips Res. Repts. Suppl. 7 (1977)
3.67   H. Haug: Festkörperprobleme, Vol. 22, 149 (Vieweg, Braunschweig 1982);
       J. P. Löwenau, S. Schmitt-Rink, H. Haug: Phys. Rev. Lett. **49**, 1511 (1982)
3.68   D. A. B. Miller, D. S. Chemla, D. J. Eilenberger, P. W. Smith, A. C. Gossard, W. T.
       Tsang: Appl. Phys. Lett. **41**, 679 (1982)
3.69   J. P. Woerdman, B. Bölger: Phys. Lett. **30A**, 164 (1969)
3.70   J. P. Woerdman: Phys. Lett. **32A**, 305 (1970)
3.71   J. P. Woerdman: Opt. Commun. **2**, 212 (1970)
3.72   S. G. Odulov, I. I. Peschko, M. S. Soskin: Ukr. Fiz. Zh. **21**, 1870 (1976)
3.73   V. L. Vinetskii, N. V. Kukhtarev, S. G. Odulov, M. S. Soskin: Sov. Phys. Techn. Phys.
       22, 729 (1977);
       V. L. Vinetskii, T. E. Zaporozkets, N. V. Kukhtarev, A. S. Matvichuck, M. S. Soskin,
       G. A. Kholodar: Ukr. Fiz. Zh. **22**, 729 (1977)
3.74   K. Jarasiunas, J. Vaitkus: Phys. Status Solidi (a) **44**, 793 (1977)
3.75   E. Gaubas, K. Jarasiunas, J. Vaitkus: Phys. Status Solidi (a) **69**, K87 (1982)
3.76   R. K. Jain, M. B. Klein: Appl. Phys. Lett. **35**, 454 (1979)
3.77   R. K. Jain, M. B. Klein, R. C. Lind: Opt. Lett. **4**, 328 (1979)
3.78   H. J. Eichler, F. Massmann: J. Appl. Phys. **53**, 3237 (1982)
3.79   F. A. Hopf, A. Tonita, T. Liepmann: Opt. Commun. **37**, 72 (1981)
3.80   E. W. van Stryland, A. L. Smirl, Th. F. Bogess, M. J. Soileau, B. S. Wherett, F. A.
       Hopf: In *Picosecond Phenomena* III, ed. by K. B. Eisenthal, R. M. Hochstrasser, W.
       Kaiser, A. Laubereau, Springer Ser. Chem. Phys., Vol. 23 (Springer, Berlin, Heidelberg
       1982) pp. 368–371
3.81   C. J. Kennedy, J. C. Matter, A. L. Smirl, H. Weichel, F. A. Hopf, S. V. Pappu, M. O.
       Scully: Phys. Rev. Lett. **32**, 419 (1974)
3.82   C. V. Shank, D. H. Auston: Phys. Rev. Lett. **34**, 479 (1975)
3.83   J. R. Lindle, S. C. Moss, A. L. Smirl: Phys. Rev. B**20**, 2401 (1979)
3.84   A. L. Smirl, Th. F. Bogess, F. A. Hopf: Opt. Commun. **34**, 463 (1980)
3.85   S. C. Moss, J. R. Lindle, H. J. Mackey, A. L. Smirl: Appl. Phys. Lett. **39**, 227 (1981)
3.86   A. L. Smirl, S. C. Moss, J. R. Lindle: Phys. Rev. B**25**, 2645 (1982)
3.87   Th. F. Bogess, A. L. Smirl, B. S. Wherett: Opt. Commun. **43**, 128 (1982)
3.88   A. L. Smirl, Th. F. Bogess, B. S. Wherett, G. P. Perryman, A. Miller: Phys. Rev. Lett.
       49, 933 (1982)
3.89   B. S. Wherett, A. L. Smirl, Th. F. Bogess: IEEE J. QE-**19**, 680 (1983)
3.90   A. L. Smirl, Th. F. Bogess, B. S. Wherett, G. P. Perryman, A. Miller: IEEE J. QE-**19**,
       690 (1983)
3.91   T. A. Wiggins, A. Salik: Appl. Phys. Lett. **25**, 438 (1974);
       T. A. Wiggins, J. A. Bellay, A. H. Carriere: Appl. Opt. **17**, 526 (1978)
3.92   R. M. Herman, C. L. Chin, E. Young: Appl. Opt. **17**, 520 (1978)
3.93   Yu. Vaitkus, E. Gaubas, K. Yarashyunas: Sov. Phys. Solid State **20**, 1824 (1978)
3.94   D. J. Ehrlich, S. R. J. Brueck, J. Y. Tsao: Appl. Phys. Lett. **41**, 630 (1982)
3.95   C. A. Hoffman, K. Jarasiunas, H. J. Gerritsen, A. V. Nurmikko: Appl. Phys. Lett. **33**,
       536 (1978);
       K. Jarasiunas, C. A. Hoffman, H. J. Gerritsen, A. V. Nurmikko: In *Picosecond
       Phenomena*, ed. by C. V. Shank, E. P. Ippen, S. L. Shapiro, Springer Ser. Chem. Phys.,
       Vol. 4 (Springer, Berlin, Heidelberg 1978) p. 327
3.96   J. Hegarty, M. D. Sturge, A. C. Gossard, W. Wiegmann: Appl. Phys. Lett. **40**, 132
       (1982)
3.97   J. G. Mendoza-Alarez, F. D. Nunes, N. B. Patel: J. Appl. Phys. **51**, 4365 (1980)
3.98   J. P. van der Ziel: IEEE J. QE-**17**, 60 (1981)
3.99   A. Olsson, C. L. Tang: Appl. Phys. Lett. **39**, 24 (1981)
3.100 C. H. Henry, R. A. Logan, K. A. Bertness: J. Appl. Phys. **52**, 4457 (1981)
3.101 Y. C. Chen, G. M. Carter: Appl. Phys. Lett. **41**, 307 (1982)
3.102 B. Jensen: IEEE J. QE-**18**, 1361 (1982)
3.103 T. A. Wiggins, J. R. Qualey: Appl. Opt. **18**, 960 (1979)

236   References

3.104 D. A. B. Miller, R. G. Harrison, A. M. Johnston, C. T. Seaton, S. D. Smith: Opt. Commun. **32**, 478 (1980)
3.105 D. R. Dean, R. J. Collins: J. Appl. Phys. **44**, 5455 (1973)
3.106 P. A. Apanasevich, A. A. Afanase'ev: Sov. Phys. Solid State **18**, 570 (1976)
3.107 K. Jarasiunas, H. J. Gerritsen: Appl. Phys. Lett. **33**, 190 (1978)
3.108 A. Borshch, M. Brodin, V. Volkov, N. Kukhtarev: Opt. Commun. **35**, 287 (1980)
3.109 A. Borshch, M. Brodin, V. Volkov, N. Krupa: Kvant. Elektr. **7**, 1557 (1980)
3.110 A. Borshch, M. Brodin, N. Orchar, S. Odulov, M. Soskin: JETP Lett. **18**, 397 (1973)
3.111 H. J. Eichler, Ch. Hartig, J. Knof: Phys. Status Solidi (a) **45**, 433 (1978)
3.112 S. G. Odulov, E. N. Salkova, L. G. Sukhoverkkova, N. M. Krokvets, G. S. Pekar, M. K. Sheinman: Ukr. Fiz. Zh. **21**, 1720 (1976)
3.113 A. Maruani, D. J. S. Chemla, E. Batifol: Solid State Commun. **33**, 805 (1980)
3.114 K. Jarasiunas, J. Vaitkus: Phys. Status Solidi (a) **23**, K19 (1974)
3.115 J. Vaitkus, Y. Vishchakas, K. Jarasiunas: Sov. J. Quant. Elektron. **5**, 1125 (1976)
3.116 R. Baltrameyunas, J. Vaitkus, K. Jarasiunas: Sov. Phys. Semicond. **10**, 572 (1976)
3.117 V. Krementskii, S. Odulov, M. Soskin: Phys. Status Solidi (a) **51**, K63 (1979) and **57**, K71 (1980)
3.118 R. K. Jain, D. G. Steel: Appl. Phys. Lett. **37**, 1 (1980)
3.119 R. K. Jain, D. G. Steel: Opt. Commun. **43**, 72 (1982)
3.120 M. A. Khan, P. W. Kruse, J. F. Ready: Opt. Lett. **5**, 261 (1980)
3.121 A. Maruani, J. L. Qudar, E. Batifol, D. S. Chemla: Phys. Rev. Lett. **41**, 1372 (1978)
3.122 A. Maruani: IEEE J. QE-16, 558 (1980)
3.123 Y. Aoyagii, Y. Segawa, S. Nomba: Phys. Rev. B**25**, 1453 (1982)
3.124 A. Ashkin, G. D. Boyd, J. M. Dziedzic, R. G. Smith, A. A. Ballman, H. J. Levinstein, K. Nassau: Appl. Phys. Lett. **9**, 72 (1966)
3.125 F. S. Chen, J. T. La Machia, D. B. Frazer: Appl. Phys. Lett. **13**, 223 (1968)
3.126 D. von der Linde, A. M. Glass: Appl. Phys. **8**, 85 (1975)
3.127 D. L. Staebler: "Ferroelectric Crystals", in *Holographic Recording Materials*, Topics Appl. Phys., Vol. 20, ed. by H. M. Smith (Springer, Berlin, Heidelberg 1977)
3.128 P. Günter: Phys. Rept. **93**, 199 (1982)
3.129 V. M. Fridkin: *Photoferroelectrics*, Springer Ser. Solid-State Sci., Vol. 9 (Springer, Berlin, Heidelberg 1979) p. 115
3.130 M. G. Moharam, T. K. Gaylord, R. Magnusson: J. Appl. Phys. **50**, 5642 (1979)
3.131 N. V. Kukhtarev, V. B. Markov, S. G. Odulov, M. S. Soskin, V. L. Vinetskii: Ferroelectrics **22**, 949 (1979)
3.132 J. Feinberg, D. Heiman, A. R. Tanguay Jr., R. W. Hellwarth: J. Appl. Phys. **51**, 1297 (1980)
3.133 J. P. Huignard, J. P. Herriau, G. Rivet, P. Günter: Opt. Lett. **5**, 102 (1980)
3.134 A. Marrakchi, J. P. Huignard, P. Günter: Appl. Phys. **24**, 131 (1981)
3.135 J. P. Huignard, A. Marrakchi: Opt. Commun. **38**, 249 (1981)
3.136 H. P. Huignard, A. Marrakchi: Opt. Lett. **6**, 622 (1981)
3.137 F. M. Küchel, H. J. Tiziani: Opt. Commun. **38**, 17 (1981)
3.138 P. Günter: Ferroelectrics **22**, 671 (1978)
3.139 P. Günter, F. Micheron: Ferroelectrics **18**, 27 (1978)
3.140 A. Krumins, P. Günter: Appl. Phys. **19**, 153 (1979)
3.141 A. E. Krumins, P. Günter: Phys. Status Solidi **55**, K185 (1979)
3.142 P. Günter, A. Krumins: Appl. Phys. **23**, 199 (1980)
3.143 A. Krumins, P. Günter: Phys. Status Solidi (a) **63**, K111 (1981)
3.144 T. K. Gaylord, T. A. Rabson, F. K. Tittel, C. R. Quick: Appl. Opt. **12**, 414 (1973)
3.145 P. Shah, T. A. Rabson, F. K. Tittel, T. K. Gaylord: Appl. Phys. Lett. **24**, 130 (1974)
3.146 N. J. Berg, B. J. Udelson, J. N. Lee: Appl. Phys. Lett. **31**, 555 (1977)
3.147 C. T. Chen, D. M. Kim, D. von der Linde: IEEE J. QE-16, 126 (1980)
3.148 J. P. Hermann, J. P. Herriau: Appl. Opt. **20**, 2173 (1981)
3.149 L. K. Lam, T. Y. Chang, J. Feinberg, R. W. Hellwarth: Opt. Lett. **6**, 475 (1981)
3.150 R. Orlowski, E. Krätzig: Solid State Commun. **27**, 1351 (1978)

3.151 E. Krätzig, F. Welz, R. Orlowski, V. Doorman, M. Rosenkranz: Solid State Commun. **34**, 817 (1980)
3.152 R. Orlowski, L. B. Boatner, E. Krätzig: Opt. Commun. **35**, 45 (1980)
3.153 R. Orlowski: Phys. Bl. **37**, 365 (1981)
3.154 V. F. Belinicher, B. I. Sturman: Sov. Phys. Usp. **23**, 199 (1980)
3.155 R. E. Aldrich, S. L. Hou, M. L. Harvill: J. Appl. Phys. **42**, 493 (1971)
3.156 S. Feinleib, D. S. Oliver: Appl. Opt. **11**, 2752 (1972)
3.157 J. P. Huignard, F. Micheron: Appl. Phys. Lett. **29**, 591 (1976)
3.158 I. P. Kaminow, E. H. Turner: "Linear Electrooptical Materials", in *Handbook of Lasers*, ed. by R. J. Pressley (Chemical Rubber, 1971)
3.159 P. Günter: Opt. Commun. **41**, 83 (1982)
3.160 P. Günter: Opt. Lett. **7**, 10 (1982)
3.161 P. Günter: Opt. Commun. **11**, 285 (1974)
3.162 S. G. Odulov: JETP Lett. **35**, 10 (1982)
3.163 H. Vorman, E. Krätzig: Solid State Commun. **49**, 843 (1984)
3.164 D. von der Linde, A. M. Glass, K. F. Rodgers: Appl. Phys. Lett. **26**, 22 (1975)
3.165 D. von der Linde, A. M. Glass, K. F. Rodgers: Appl. Phys. Lett. **25**, 155 (1974)
3.166 D. von der Linde, A. M. Glass, K. F. Rodgers: J. Appl. Phys. **47**, 217 (1976)
3.167 G. C. Valley: IEEE J. QE-**19**, 1637 (1983)
3.168 L. D. Landau, E. M. Lifshitz: *Fluid Mechanics* (Pergamon, Oxford 1963) Chap. 51
3.169 D. W. Pohl, S. E. Schwarz, V. Irniger: Phys. Rev. Lett. **31**, 32 (1973)
3.170 W. Chan, P. S. Pershan: Phys. Rev. Lett. **39**, 1368 (1977)
3.171 M. Chester: Phys. Rev. **131**, 2013 (1963);
       E. W. Prohofsky, J. A. Krummhansl: Phys. Rev. **133**, A1411 (1964);
       R. A. Guyer, J. A. Krummhansl: Phys. Rev. Lett. **148**, 766 (1966)
3.172 D. W. Pohl, V. Irniger: Phys. Rev. Lett. **36**, 480 (1976)
3.173 D. W. Pohl: Solid State Commun. **23**, 447 (1977)
3.174 R. K. Wehner, R. Klein: Physica **62**, 5161 (1972)
3.175 D. W. Pohl: Phys. Rev. Lett. **43**, 143 (1979)
3.176 H. Hervet, W. Urbach, F. Rondelerz: J. Chem. Phys. **68**, 2725 (1978)
3.177 K. Thyagarajan, P. Lallemand: Opt. Commun. **26**, 54 (1978)
3.178 D. W. Pohl: Phys. Lett. **77A**, 53 (1980)
3.179 P. Y. Key, R. G. Harrison, V. I. Little, J. Katzenstein: IEEE J. QE-**6**, 641 (1970)
3.180 R. G. Harrison, P. Y. Key, V. I. Little: Proc. R. Soc. Lond. A. **334**, 193 (1973)
3.181 R. G. Harrison, P. Y. Key, V. I. Little: Proc. R. Soc. Lond. A. **334**, 215 (1973)
3.182 S. D. Durbin, S. M. Arakelian, Y. R. Shen: Opt. Lett. **7**, 145 (1982)
3.183 D. Veletskas, I. Kapturauskas, R. Baltrameyunas: Sov. Phys. Tech. Phys. **27**, 263 (1982)
3.184 S. A. Akhmanow, R. V. Khokhlov, A. P. Sukhornkov: "Self-Focussing, Self-Defocussing and Self-Modulation of Laser Beams", in *Laser Handbook*, Vol. 2, ed. by F. T. Arecchi, E. O. Schulz-Dubois (North-Holland, Amsterdam 1982) p. 1151
3.185 P. P. Ho, R. R. Alfano: Phys. Rev. A**20**, 2170 (1979)
3.186 T. Y. Chang: Opt. Eng. **20**, 220 (1981)
3.187 E. Gaubas, K. Jarasunas, J. Vaitkus: Sov. Phys. Collect. **21**, 56 (1981)
3.188 J. Vaitkus, M. Pyatravskas, K. Yarashyunas: Sov. Phys. Tech. Semic. **16**, 650 (1982)
3.189 D. J. Hagan, H. A. MacKenzie, H. A. AlAttar, W. Y. Firth: Opt. Lett. **10**, 187 (1985)
3.190 Yu. Vaitkus, E. Gaubas, E. V. Ivakin, S. I. Mironmko, A. S. Rubanov, K. Yarashyunas: Sov. Quant. Electr. **13**, 856 (1983)
3.191 H. Kalt, V. G. Lyssenko, R. Renner, C. Klingshirn: Solid State Comm. **51**, 675 (1984); JOSA B**2**, 1188 (1985)
3.192 H. Kalt, V. G. Lyssenko, K. Bohnert, C. Klingshirn: J. Luminesc. **31**, **32**, 861 (1985)
3.193 J. Vaitkus, H. J. Gerritsen, K. Jarashunas, R. Baltrameunas: JOSA **70**, 616 (1980)
3.194 I. Rückmann, K. Jarasunas, E. Gaubas: Phys. Status Solidi (b) **128**, 627 (1985)

Chapter 4

4.1   H. P. Yuen, J. H. Shapiro: Opt. Lett. **4**, 334 (1979)
4.2   T. K. Gaylord, M. G. Moharam: Appl. Phys. **B28**, 1 (1982)
4.3   R. Petit (ed.): *Electromagnetic Theory of Gratings*, Topics Curr. Phys. Vol. 22 (Springer, Berlin, Heidelberg 1980)
4.4   L. Solymar, D. J. Cooke: *Volume Holography and Volume Grating* (Academic, London 1981)
4.5   B. Benlarbi, P. St. J. Rusell, L. Solymar: Appl. Phys. **B28**, 63 (1982)
4.6   R. Magnusson, T. K. Gaylord: Opt. Commun. **28**, 1 (1979)
4.7   E. G. Loewen, M. Neviere, D. Maystre: Appl. Opt. **16**, 2711 (1977)
4.8   M. J. Hayford: Photonics Spectra (April 1982)
4.9   H. Kogelnik: Bell Syst. Tech. J. **48**, 2909 (1969)
4.10 H. J. Eichler: *Adv. Solid State Phys.* **18**, 241 (Vieweg, Braunschweig 1978)
4.11 P. St. J. Rusell, L. Solymar: Appl. Phys. **22**, 335 (1980)
4.12 V. L. Vinetskii, N. V. Kukhtarev, S. G. Odulov, M. S. Soskin: Sov. Phys. Vsp. **22**, 742 (1979)
4.13 R. H. Enns, S. S. Rangnekar: Can. J. Phys. **52**, 99 and 562 (1974)
4.14 H. J. Eichler, G. Enterlein, J. Munschau, H. Stahl: Z. Angew. Phys. **31**, 1 (1971)
4.15 D. L. Staebler, J. J. Amodei: J. Appl. Phys. **43**, 1042 (1972)
4.16 V. L. Vinetskii, N. V. Kukhtarev, M. S. Soskin: Sov. J. Quant. Electr. **7**, 230 (1977)
4.17 N. Bloembergen: In *Laser Spectroscopy IV*, ed. by H. Walther and K. W. Rothe, Springer Ser. Opt. Sci., Vol. 21 (Springer, Berlin, Heidelberg 1979) p. 340
4.18 N. Bloembergen: *Nonlinear Optics* (Benjamin, New York 1965);
      E. Yablonowitch, N. Bloembergen, J. J. Wynne: Phys. Rev. **B10**, 4447 (1974);
      M. D. Levenson, N. Bloembergen: Phys. Rev. **B10**, 4447 (1974)
4.19 P. D. Maker, R. W. Terhune: Phys. Rev. **A137**, 801 (1965)
4.20 A. Yariv: IEEE J. **QE-14**, 650 (1978); and **QE-15** 524 (1979);
      J. O. White, A. Yariv: Opt. Eng. **21**, 224 (1982)
4.21 R. L. Abrams, R. C. Lind: Opt. Lett. **2**, 94 (1978)
4.22 R. G. Caro, M. C. Gower: IEEE J. **QE-18**, 1376 (1982)
4.23 D. G. Steel, R. C. Lind, J. F. Lam, C. R. Giuliano: Appl. Phys. Lett. **35**, 376 (1979)
4.24 D. K. Saldin, T. Wilson, L. Solymar: J. Opt. Soc. Am. **72**, 1179 (1982)
4.25 M. Ducloy: *Adv. Solid State Phys.* 22, 35 (Vieweg, Braunschweig 1982)
4.26 J. H. Marburger, J. F. Lam: Appl. Phys. Lett. **35**, 249 (1979)
4.27 J.-C. Diels, W.-C. Wang: Appl. Phys. **B26**, 105 (1981)
4.28 R. C. Shockley: Opt. Commun. **38**, 221 (1981)
4.29 Y. Silberberg, I. Bar-Joseph: IEEE J. **QE-17**, 1967 (1981)
4.30 R. C. Shockley: Appl. Phys. Lett. **40**, 930 (1982)
4.31 P. F. Liao, D. M. Bloom, N. P. Economou: Opt. Lett. **2**, 58 (1978)
4.32 R. C. Lind, D. G. Steel, M. B. Klein, R. L. Abrams, C. R. Giuliano: Appl. Phys. Lett. **34**, 457 (1979)
4.33 D. G. Steel, J. F. Lam: Opt. Lett. **4**, 363 (1979)
4.34 Y. Silberberg, I. Bar-Joseph: Opt. Commun. **39**, 265 (1981)
4.35 J. L. Ferrier, Z. Wu, X. Nguyen Phu, G. Rivoire: Opt. Commun. **41**, 207 (1982)
4.36 N. S. Vorobiev, I. S. Ruddock, R. Illingworth: Opt. Commun. **41**, 216 (1982)
4.37 R. K. Jain, M. B. Klein: Appl. Phys. Lett. **35**, 454 (1979)
4.38 N. Kukhtarev, S. Odoulov: Opt. Commun. **32**, 183 (1980)
4.39 P. Günter: Phys. Rept. **93**, 199 (1982)
4.40 Y. H. Ja: Opt. Commun. **41**, 159 (1982)
4.41 J. Feinberg: Opt. Lett. **7**, 486 (1982)
4.42 M. Cronin-Goulomb, B. Fischer, J. O. White, A. Yariv: IEEE J. **QE-20**, 12 (1981)
4.43 D. E. Watkins, J. F. Figueira, S. J. Thomas: Opt. Lett. **5**, 169 (1980)
4.44 K. Ujihara: Opt. Commun. **42**, 1 (1982)
4.45 B. Y. Zel'dovich, N. F. Pilipetsky, V. V. Shkunov: *Principles of Phase Conjugation*, Springer Ser. Opt. Sci., Vol. 42 (Springer, Berlin, Heidelberg 1985)

4.46   D. M. Pepper, R. L. Abrams: Opt. Lett. **3**, 212 (1978)
4.47   G. P. Agrawal, C. Flytzanis, R. Frey, F. Pradere: Appl. Phys. Lett. **38**, 492 (1981)
4.48   J.-L. Oudar, Y. R. Shen: Phys. Rev. A**22**, 1141 (1980)
4.49   D. G. Steel, R. C. Lind, J. F. Lam: Phys. Ref. A**23**, 2513 (1981)
4.50   Peixian Ye, Y. R. Shen: Phys. Rev. A**25**, 2183 (1982)
4.51   J. F. Lam, R. L. Abrams: Phys. Rev. A**26**, 1539 (1982)
4.52   A. E. Siegmann: Appl. Phys. Lett. **30**, 21 (1977);
        R. Trebino, A. E. Siegmann: Appl. Phys. B**28**, 250 (1982)
4.53   H. J. Eichler: Optica Acta **24**, 631 (1977)
4.54   V. N. Mahajan, J. D. Gaskill: J. Appl. Phys. **45**, 2799 (1974)
4.55   M. D. Levenson: Physics Today 3, 44 (1977);
        W. M. Tolles, J. W. Nibler, R. McDonald, A. B. Harvey: Appl. Spectr. **31**, 253 (1977)
4.56   C. Flytzanis: In *Quantum Electronics*, ed. by H. Rubin and C. L. Tang (Academic, New
        York 1975) Vol. 1
4.57   Y. R. Shen: Rev. Mod. Phys. **48**, 1 (1976)

Chapter 5

5.1   H. J. Eichler, Ch. Hartig, J. Knof: Phys. Status Solidi (a) **45**, 433 (1978)
5.2a)   D. W. Phillion, D. J. Kuizenga, A. E. Siegmann: Appl. Phys. Lett. **27**, 85 (1975);
5.2b)   A. E. Siegmann: Appl. Phys. Lett. **20**, 21 (1977)
5.3   T. Yajima: Opt. Commun. **14**, 378 (1975);
        T. Yajima, H. Souma, Y. Ishida: Opt. Commun. **18**, 150 (1976); Phys. Rev. A**17**, 309
        and 324 (1978);
        T. Yajima: J. Phys. Soc. Japan **44**, 948 (1978)
5.4   H. Eichler, G. Salje, H. Stahl: J. Appl. Phys. **44**, 5383 (1973);
        H. Eichler, G. Enterlein, P. Glozbach, J. Munschau, H. Stahl: Appl. Opt. **11**, 372
        (1972)
5.5   D. W. Pohl, S. E. Schwarz, V. Irniger: Appl. Phys. Lett. **31**, 32 (1973)
5.6   H. E. Jackson, C. T. Walker: Phys. Rev. B**3**, 1428 (1971);
        T. F. McNelly, S. J. Rogers, D. J. Channin, R. J. Rollefson, W. M. Gouban, G. E.
        Schmidt, J. A. Krumhansl, R. O. Pohl: Appl. Phys. Lett. **24**, 100 (1970)
5.7   A comprehensive list of references on the extended second-sound literature can be
        found in H. Beck, P. F. Meier, A. Thellung: Phys. Status Solidi A**24**, 11 (1974), and in
        the papers of the previous reference
5.8a)   D. W. Pohl, S. E. Schwarz: Phys. Rev. B**7**, 2735 (1973);
5.8b)   D. W. Pohl: IBM J. Res. Develop. **23**, 604 (1979)
5.9   See, e.g., L. D. Landau, E. M. Lifshitz: *Statistical Physics* (Pergamon, Oxford, 1980),
        67
5.10   H. J. Eichler, J. Knof: Appl. Phys. **13**, 209 (1977)
5.11   D. W. Pohl: Proc. 3rd Intern. Conf. on Light Scattering in Solids, ed. by M. Balkanski,
        R. C. C. Leite, and S. P. S. Porto (Flammarion, Paris 1976);
        D. W. Pohl, V. Irniger: Phys. Rev. Lett. **36**, 480 (1976)
5.12   W. Urbach, H. Hervet, F. Rondelez: Mol. Cryst. Liq. Cryst. **46**, 209 (1978);
        F. Rondelez, W. Urbach, H. Hervet: Phys. Rev. Lett. **41**, 1058 (1978)
5.13   D. W. Pohl: Appl. Phys. Lett. **43**, 143 (1979)
5.14   See, e.g., the reviews of R. O. Pohl and G. L. Salinger: Annals N.Y. Acad. Sci. **279**, 150
        (1976), or S. Hunklinger, W. Arnold: in *Physical Acoustics*, Vol. 12 (Academic, New
        York 1976)
5.15   C. Cohen: Phys. Rev. B**13**, 866 (1976)
5.16   P. A. Fleury, K. B. Lyons: Phys. Rev. Lett. **36**, 1188 (1976)
5.17   J. A. Cowen, C. Allain, P. Lallemand: J. Physique Lett. **37**, 313 (1976)
5.18   R. D. Mountain: J. Res. Nat. Bur. Stand. **70A**, 207 (1966) and **72A**, 95 (1968)
5.19   W. Chan, P. S. Pershan: Phys. Rev. Lett. **39**, 1368 (1977)

240   References

5.20   See, e.g., T. Riste (ed.): *Fluctuations, Instabilities, and Phase Transitions* (Plenum, New York 1975)
5.21   C. Allain, H. Z. Cummins, P. Lallemand: J. Physique Lett. **39**, L475 (1978)
5.22   J. P. Boon, C. Allain, P. Lallemand: Phys. Rev. Lett. **43**, 199 (1979)
5.23   W. Marine, J. Marfaing, F. Salvan: J. Physique Lett. **44**, L271 (1983)
5.24   K. Thyagarayan, P. Lallemand: Opt. Commun. **26**, 54 (1978)
5.25   D. W. Pohl: Phys. Lett. **77A**, 53 (1980)
5.26   A. W. Lowen, O. K. Rice: Trans. Faraday Soc. **59**, 2723 (1963)
5.27   E. Guelari, A. F. Collings, R. L. Schmidt, C. J. Pings: J. Chem. Phys. **56**, 6169 (1972)
5.28   H. Hervet, W. Urbach, F. Rondelez: J. Chem. Phys. **68**, 2725 (1978)
5.29   F. Rondelez: Solid State Commun. **14**, 815 (1974)
5.30   J. F. Wesson, H. Takezoe, H. Yu: J. Appl. Phys. **53**, 6513 (1982)
5.31   H. Hervet, L. Leger, F. Rondelez: Phys. Rev. Lett. **42**, 1681 (1979)
5.32   N. Nemoto, M. R. Landry, I. Noh, H. Yu: Polym. Commun. **25**, 141 (1984)
5.33   D. G. Miles, P. D. Lamb, K. W. Rhee, C. S. Johnson: J. Phys. Chem. **87**, 4815 (1983);
       K. W. Rhee, D. A. Gabriel, C. S. Johnson: J. Phys. Chem. **88**, 4010 (1984);
       K. W. Rhee, J. Shibata, A. Barish, D. A. Gabriel, C. S. Johnson: **88**, 394
5.34   H. Kim, T. Chang, H. Yu: J. Phys. Chem., **88**, 3946
5.35   J. A. Wesson, I. Noh, T. Kitano, H. Yu: Macromolec. **17**, 783 (1984)
5.36   A. Korpel, R. Adler, B. Alpiner: Appl. Phys. Lett. **5**, 86 (1964)
5.37   D. E. Caddes, C. F. Quate, C. D. W. Wilkinson: In *Proc. Symp. on Modern Optics* (Polytechnic Press, Brooklyn 1967) p. 219
5.38   R. E. Lee, R. M. White: Appl. Phys. Lett. **12**, 12 (1968)
5.39   D. C. Auth: Appl. Phys. Lett. **16**, 521 (1970)
5.40   G. Cachier: Appl. Phys. Lett. **17**, 419 (1970)
5.41   H. Eichler, H. Stahl: Opt. Commun. **6**, 239 (1972)
5.42   H. Eichler, H. Stahl: J. Appl. Phys. **44**, 3429 (1973)
5.43   E. V. Ivakin, A. M. Lazaruk, I. P. Petrovich, A. S. Rubanov: Kvant. Elektronika **4**, 2421 (1977)
5.44   F. V. Bunkin, M. I. Tribelskil: Sov. Phys. Usp. **23**, 105 (1980)
5.45   J. R. Salcedo, A. E. Siegman: IEEE J. QE-**15**, 250 (1979)
5.46   K. A. Nelson, M. D. Fayer: J. Chem. Phys. **72**, 5205 (1980)
5.47   K. A. Nelson, R. J. D. Miller, D. R. Lutz, M. D. Fayer: J. Appl. Phys. **53**, 1144 (1982)
5.48   K. A. Nelson: J. Appl. Phys. **53**, 6060 (1982)
5.49   R. J. Miller, R. Casalegno, K. A. Nelson, M. D. Fayer: Chem. Phys. **72**, 371 (1982)
5.50   K. A. Nelson, R. Casalegno, R. J. D. Miller, M. D. Fayer: J. Chem. Phys. **77**, 1144 (1982)
5.51   H. Stahl: „Untersuchungen an laserinduzierten Ultraschallwellen", Dissertation, Technische Universität Berlin (1973)
5.52   P. G. De Gennes: J. Physique Lett. **38**, 1 (1977)
5.53   M. Fermigier, E. Guyon, P. Jenffer, L. Petit: Appl. Phys. Lett. **36**, 361 (1980)
5.54   M. Fermigier, P. Jenffer, J. C. Charmet, E. Guyon: J. Physique Lett. **41**, 519 (1980)
5.55   M. Fermigier, M. Cloitre, E. Guyon, P. Jenffer: J. Mecan. Theor. et Appl. **1**, 123 (1982)
5.56   F. Durst, A. Melling, J. H. Whitelaw: *Principles and Practice of Laser Doppler Anemometry* (Academic, London 1976)
5.57   G. Enterlein: Diplomarbeit, Optisches Institut, Technische Universität Berlin (1971)
5.58   N. Wiese: Studienarbeit, Optisches Institut, Technische Universität Berlin (1982)
5.59   D. H. Auston, C. V. Shank, P. Lefur: Phys. Rev. Lett. **35**, 1035 (1976)
5.60   B. Sermage, H. J. Eichler, J. P. Heritage, R. J. Nelson, N. K. Dutta: Appl. Phys. Lett. **42**, 259 (1983)
5.61   H. J. Eichler, F. Massmann: J. Appl. Phys. **53**, 3237 (1982)
5.62   J. P. Woerdman: Philips Res. Rep. Suppl. **7** (1971)
5.63   K. Jarasiunas, J. Vaitkus: Phys. Stat. Sol. A**44**, 793 (1977)
5.64   S. C. Moss, J. R. Lindle, H. J. Mackey, A. L. Smirl: Appl. Phys. Lett. **39**, 227 (1981)
5.65   K. Jarasiunas, H. J. Gerritsen: Appl. Phys. Lett. **33**, 190 (1978)

5.66   I. Broser, R. Broser, M. Rosenzweig: In *Landolt-Börnstein*, Vol. 176 (Springer, Berlin, Heidelberg 1982) p. 190
5.67   C. A. Hoffmann, K. Jarasiunas, A. V. Nurmikko, H. J. Gerritsen: Appl. Phys. Lett. **33**, 536 (1980)
5.68   C. A. Hoffmann, H. J. Gerritsen: J. Appl. Phys. **51**, 1603 (1980)
5.69   R. C. Powell, G. Blasse: "Energy Transfer in Concentrated Systems" in *Structure and Bonding*, Vol. 42 (Springer, Berlin, Heidelberg 1980)
5.70   M. D. Fayer: Ann. Rev. Phys. Chem. **33**, 63–87 (1982)
5.71   R. J. D. Miller, M. Pierre, M. D. Fayer: J. Chem. Phys. **78**, 5138 (1983)
5.72   S. K. Lyo: Phys. Rev. B3, 3331 (1971)
5.73   P. W. Anderson: Phys. Rev. **109**, 1492 (1958)
5.74   H. J. Eichler, J. Eichler, J. Knof, Ch. Noack: Phys. Status Solidi (a) **52**, 481 (1979)
5.75   D. S. Hamilton, D. Heiman, J. Feinberg, R. W. Hellwarth: Opt. Lett. **4**, 124 (1979)
5.76   P. F. Liao, L. M. Humphrey, D. M. Bloom, S. Geschwind: Phys. Rev. B20, 4145 (1979)
5.77   Ch. M. Lawson, R. C. Powell, W. K. Zwicker: Phys. Rev. Lett. **46**, 1020 (1981)
5.78   J. R. Salcedo, A. E. Siegman, D. D. Dlott, M. D. Fayer: Phys. Rev. Lett. **41**, 131 (1978)
5.79   P. Günter, F. Micheron: Ferroelectrics **18**, 27 (1978)
5.80   R. Orlowski, E. Krätzig: Solid State Commun. **27**, 1351 (1978)
5.81   A. Krumins, P. Günter: Appl. Phys. **19**, 153 (1979)
5.82   P. Günter: Physics Rept. **93**, 199–299 (1982)
5.83   F. Rondelez, H. Hervet, W. Urbach: Chem. Phys. Lett. **53**, 138 (1978)
5.84   D. M. Burland, G. C. Bjorklund, D. C. Alvarez: J. Am. Chem. Soc. **102**, 7117 (1980)
5.85   G. C. Bjorklund, D. M. Burland, D. C. Alvarez: J. Chem. Phys. **73**, 4321 (1980)
5.86   D. M. Burland, Chr. Bräuchle: J. Chem. Phys. **76**, 4502 (1982)
5.87   D. M. Burland: Acc. Chem. Res. **16**, 218 (1983)
5.88   Chr. Bräuchle, D. M. Burland: Angew. Chem. Int. Ed. Engl. **22**, 582 (1983)
5.89   J. R. Andrews, R. M. Hochstrasser: Chem. Phys. Lett. **76**, 207 (1980)
5.90   H. J. Gerritsen, M. E. Heller: J. Appl. Phys. **38**, 2054 (1967)
5.91   M. A. Cutter, P. Y. Key, V. I. Little: Appl. Opt. **13**, 1399 (1974); and **15**, 2954 (1976)
5.92   W. Marine, J. Marfaing, F. Salvan: J. Physique Lett. **44**, 271 (1983)
5.93   A. L. Dalisa, W. K. Zwicker, D. J. DeBitetto, P. Harnack: Appl. Phys. Lett. **17**, 208 (1970)
5.94   R. M. Osgood, Jr., A. Sanchez-Rubio, D. J. Ehrlich, V. Daneu: Appl. Phys. Lett. **40**, 391 (1982)
5.95   D. V. Podlenik, H. H. Gilgen, R. M. Osgood, A. Sanchez: Appl. Phys. Lett. **43**, 1083 (1983)
5.96   M. Birnbaum: J. Appl. Phys. **36**, 3688 (1965)
5.97   M. Siegrist, G. Kaech, F. K. Kneubühl: Appl. Phys. **2**, 45 (1973)
5.98   J. F. Young, J. E. Sipe, J. S. Preston, H. M. vanDriel: Appl. Phys. Lett. **41**, 261 (1982)
5.99   F. Keilmann, Y. H. Bai: Appl. Phys. A29, 9 (1982)
5.100 Zhou Guosheng, P. M. Fauchet, A. E. Siegman: Phys. Rev. B26, 5366 (1982)
5.101 S. R. J. Brueck, D. J. Ehrlich: Phys. Rev. Lett. **48**, 1678 (1982)
5.102 D. J. Ehrlich, S. R. J. Brueck, J. Y. Tsao: Appl. Phys. Lett. **41**, 630 (1982)
5.103 F. Keilmann: In *Surface Studies with Lasers*, ed. by F. R. Aussenegg, A. Leitner, M. E. Lippitsch, Springer Ser. Chem. Phys., Vol. 33 (Springer, Berlin, Heidelberg 1983)
5.104 V. Butkus: Sov. Phys. Coll. **23**, No. 6 (1983)
5.105 S. Komuro, Y. Aoyagi, Y. Segawa, S. Namba, A. Masuyama, H. Okamoto, Y. Hamakawa: Appl. Phys. Lett. **42**, 79 (1983)
5.106 T. Miyoshi, Y. Aoyagi, Y. Segawa, S. Namba, H. Okamoto, Y. Hamakawa: Jap. J. Phys. **22**, 886 (1983)
5.107 J. Vaitkus: Sov. Phys. Coll. **25**, No. 4 (1985)

Chapter 6

6.1   R. J. Collier, Ch. B. Burckhardt, L. H. Lin: *Optical Holography* (Academic, New York 1971)
6.2   L. Solymar, D. J. Cooke: *Volume Holography and Volume Gratings* (Academic, London 1981)
6.3   H. M. Smith: *Principles of Holography* (Wiley, New York 1975)
6.4   W. T. Cathey: *Optical Information Processing and Holography* (Wiley, New York 1974)
6.5   M. Françon: *Holographie* (Springer, Berlin, Heidelberg 1972)
6.6   G. W. Stroke: *An Introduction to Coherent Optics and Holography* (Academic, New York 1969)
6.7   H. M. Smith: *Holographic Recording Materials*, Topics Appl. Phys., Vol. 20 (Springer, Berlin, Heidelberg 1977)
6.8   H. Kogelnik: Bell Syst. Tech. J. **48**, 2909 (1969)
6.9   P. Günter: Physics Rept. **93**, 199 (1982)
6.10 H. Eichler: Z. Angew. Phys. **31**, 1 (1971)
6.11 J. W. Goodmann: *Introduction to Fourier Optics* (McGraw-Hill, New York 1968)
6.12 J. P. Huignard, J. P. Herriau, T. Valentin: Appl. Opt. **16**, 2796 (1977)
6.13 J. P. Huignard, J. P. Herriau: Appl. Opt. **16**, 180 (1977)
6.14 A. Marrakchi, J. P. Huignard, J. P. Herriau: Opt. Commun. **34**, 15 (1980)
6.15 P. St. J. Russell: Appl. Phys. **B26**, 37 and 89 (1981)
6.16 J. C. Dainty: *Laser Speckle and Related Phenomena*, 2nd ed., Topics Appl. Phys., Vol. 9 (Springer, Berlin, Heidelberg 1984)
6.17 J. P. Huignard, J. P. Herriau, L. Pichon, A. Marrakchi: Opt. Lett. **5**, 436 (1980)
6.18 F. M. Küchel, H. J. Tiziani: Opt. Commun. **38**, 17 (1981)
6.19 J. Feinberg: Opt. Lett. **5**, 330 (1980)
6.20 J. P. Huignard, J. P. Herriau: "Recent Advances in Holography", SPIE Proc., Vol. 215 (1980) p. 178
6.21 D. von der Linde, A. M. Glass: Appl. Phys. **8**, 88 (1975)
6.22 O. N. Tufte, D. Chen: IEEE Spectrum **10**, 26 (1973)
6.23 H. Kurz: Philips Tech. Rev. **37**, 109 (1977)
6.24 H. Kurz: Optica Acta **24**, 463 (1977)
6.25 J. D. Zook: Appl. Opt. **13**, 875 (1974)
6.26 M. R. Latta, R. V. Pole: Appl. Opt. **18**, 2418 (1979)
6.27 A. M. P. P. Leite, O. D. D. Soares, E. A. Ash: Optics and Acoustics **2**, 45 (1978)
6.28 E. A. Ash, E. Seatford, O. Soares, K. S. Remington: Appl. Phys. Lett. **24**, 207 (1974)
6.29 Yu, A. Bykovoky, A. V. Makovkin, V. L. Smirnov: Opt. Spectrosc. **37**, 579 (1984)
6.30 H. Nishihara, S. Inohara, T. Suhara, J. Koyama: IEEE J. QE-**11**, 794 (1975)
6.31 G. Goldmann, H. H. Witte: Opt. & Quant. Electron. **9**, 75 (1977)
6.32 H. J. Gerritsen, E. G. Ramberg, S. Freemann: *Symposium on Modern Optics* (Polytechnic Institute of Brooklyn 1967) p. 109
6.33 G. Martin, R. W. Hellwarth: Appl. Phys. Lett. **34**, 371 (1979)
6.34 P. Günter: Ferroelectrics **40**, 43 (1982)
6.35 F. Laeri, T. Tschudi, H. Albers: Opt. Commun. **47**, 387 (1983)
6.36 A. E. Krumins, P. Günter: Sov. J. Quantum Electron. **10**, 681 (1980)
6.37 P. Refregier, L. Solymar, H. Rajbenbach, J.-P. Huignard: Electron. Lett. **20**, 656 (1984)
6.38 V. Markov, S. Odulov, M. Soskin: Optics and Laser Techn. 95 (April 1979)
6.39 J. P. Huignard, J. P. Herriau, F. Micheron: Appl. Phys. Lett. **26**, 256 (1975)
6.40 J. P. Huignard, F. Micheron, E. Spitz: in *Optical Properties of Solids: New Developments*, ed. by B. O. Scraphin (North Holland, Amsterdam 1976) pp. 851–922
6.41 J. P. Huignard, J. P. Herriau, F. Micheron: Ferroelectrics **11**, 393 (1976)
6.42 C. S. Weaver, J. N. Goodmann: Appl. Opt. **5**, 1248 (1966)
6.43 J. O. White, A. Yariv: Appl. Phys. Lett. **37**, 5 (1980)
6.44 D. M. Pepper, J. Au Yeung, D. Fekete, A. Yariv: Opt. Lett. **3**, 7 (1978)

6.45 N. D. Ustinov, L. N. Mateev: Soc. J. Quant. Electron. **7**, 1483 (1977)
6.46 L. Pichon, J. P. Huignard: Opt. Commun. **36**, 277 (1981)
6.47 R. W. Hellwarth: J. Opt. Soc. Am. **67**, 1 (1977)
6.48 A. Yariv: IEEE J. QE-**14**, 650 (1978)
6.49 C. R. Giuliano: Phys. Today 27–34 (April 1981)
6.50 B. Ya. Zel'dovich, N. F. Pilipetsky, V. V. Shkunov: *Principles of Phase Conjugation*, Springer Ser. Opt. Sci., Vol. 42 (Springer, Berlin, Heidelberg 1985);
      B. Ya. Zel'dovich, N. F. Pilipetskii, V. V. Shkunov: Sov. Phys. Usp. **25**, 713 (1982)
6.51 M. Ducloy: In *Festkörperprobleme*, Vol. 22 (Vieweg, Braunschweig 1982);
      D. M. Pepper (ed.) Special issue on Nonlinear Optical Phase Conjugation, Optical Eng. **21**, (February 1982)
6.52 R. A. Fischer (ed.) *Optical Phase Conjugation* (Academic, New York 1983)
6.53 H. Kogelnik: Bell Syst. Tech. J. **44**, 2451 (1965);
      E. N. Leith, J. Upatnieks: J. Opt. Soc. Am. **56**, 523 (1966);
      J. P. Woerdman: Opt. Commun. **2**, 212 (1971);
      B. Ya. Zel'dovich, V. I. Ragul'skii, F. S. Faizullov: Sov. Phys. JETP **15**, 109 (1972)
6.54 W. T. Cathey: Proc. IEEE **56**, 340 (1968);
      W. T. Cathey, C. L. Hayes, W. C. Davis, V. F. Pizzuro: Appl. Opt. **9**, 701 (1970)
6.55 M. D. Levenson, K. M. Jonson, V. C. Hanchett, K. Chiang: J. Opt. Soc. Am. **71**, 737 (1981)
6.56 M. D. Levenson: Opt. Lett. **5**, 182 (1980)
6.57 J. Hubbard: J. Opt. Soc. Am. **71**, 1029 (1981)
6.58 J. Feinberg, R. Hellwarth: Opt. Lett. **5**, 519 (1980) and Erratum, Opt. Lett. **6**, 257
6.59 B. Fischer, M. Cronin-Golomb, J. O. White, A. Yariv, R. Neurgaonkar: Appl. Phys. Lett. **40**, 863 (1982)
6.60 P. Günter: Opt. Lett. **7**, 10 (1982)
6.61 A. Yariv: Appl. Phys. Lett. **28**, 88 (1976)
6.62 G. C. Valley, G. J. Dunning: Opt. Lett. **9**, 513 (1983)
6.63 T. R. O'Meara: Opt. Eng. **21**, 243 (1982)
6.64 P. Günter, E. Voit, M. Z. Zha, H. Albers: Opt. Commun. **55**, 210 (1985)
6.65 K. R. Mac Donald, J. Feinberg: J. Opt. Soc. Am. **73**, 548 (1983)
6.66 S. M. Jensen, R. W. Hellwarth: Appl. Phys. Lett. **33**, 404 (1978)
6.67 J. O. White, M. Cronin-Golomb, B. Fischer, A. Yariv: Appl. Phys. Lett. **40**, 450 (1982)
6.68 M. Cronin-Golomb, B. Fischer, J. O. White, A. Yariv: Appl. Phys. Lett. **41**, 689 (1982)
6.69 J. Feinberg: Opt. Lett. **7**, 486 (1982)
6.70 J. Feinberg: J. Opt. Soc. Am. **72**, 46 (1982)
6.71 D. M. Pepper, D. Fekete, A. Yariv: Appl. Phys. Lett. **33**, 41 (1978)
6.72 J. Feinberg, G. D. Bacher: Opt. Lett. **9**, 420 (1984)
6.73 B. J. Feldmann, I. J. Bigio, R. A. Fischer, C. R. Phipps Jr., D. E. Watkins, S. J. Thomas: Los Alamos Science **3** (Fall 1982)
6.74 K. R. MacDonald, J. Feinberg: J. Opt. Soc. Am. **73**, 548 (1983)
6.75 A. Litvinenko, S. Odulov: Opt. Lett. **9**, 68 (1984)
6.76 J. F. Lam, W. P. Brown: Opt. Lett. **5**, 61 (1980)

Chapter 7

7.1 C. L. Tang, H. Statz, G. DeMars: J. Appl. Phys. **34**, 2289 (1963)
7.2 S. E. Harris, O. P. McDuff: IEEE J. QE-**2**, 47 (1966)
7.3 H. Eichler, P. Glozbach, B. Kluzowski: Z. Angew. Phys. **28**, 303–306 (1970)
7.4 H. G. Danielmeyer: J. Appl. Phys. **42**, 3125–3132 (1971)
7.5 H. J. Eichler, J. Eichler: J. Appl. Phys. **45**, 4950 (1974)
7.6 J. B. Hambenne, M. Sargent III: IEEE J. QE-**11**, 90 (1975)
7.7 M. Sargent III: Appl. Phys. **9**, 127–141 (1976)
7.8 D. Kühlke, R. Horak: Opt. Quant. Elect. **11**, 485–495 (1979)

244   References

7.9   E. Kyrölä, R. Salomaa: Appl. Phys. **20**, 339–344 (1979)
7.10   A. Bambini, R. Vallauri, R. Karamaliev: Phys. Rev. **A19**, 1673 (1979)
7.11   H. Kressel, J. K. Butler: Semiconductor Laser and Heterojunction LED's (Academic, New York, 1977)
7.12   B. W. Haki: J. Appl. Phys. **46**, 292 (1975)
7.13   H. Kawaguchi, K. Takahei: IEEE J. QE-**16**, 706 (1980)
7.14   H. Eichler, G. Schick, W. Wiesemann: IEEE J. QE-**11**, 168 (1975)
7.15   D. Dammasch, H. J. Eichler, G. Schick: Rev. Bras. de Fisica **10**, 239 (1980)
7.16   C. G. Aminoff, M. Kaivola: Appl. Phys. **B26**, 133 (1981)
7.17   C. G. Aminoff, M. Kaivola: Opt. Commun. **37**, 133 (1981)
7.18   R. L. Fork, B. I. Greene, C. V. Shank: Appl. Phys. Lett. **38**, 671 (1981)
7.19   J.-C. Diels, J. J. Fontaine, I. C. McMichael, C. Y. Wang: Appl. Phys. **B28**, 172 (1982)
7.20   R. G. Harrison, P. Key, V. I. Little, G. Magyar, J. Katzenstein: Appl. Phys. Lett. **13**, 253 (1968)
7.21   J. Katzenstein, G. Magyar, A. C. Selden: Opto-Elec. **1**, 13 (1969)
7.22   H. Kigelnick, C. V. Shank: Appl. Phys. Lett. **18**, 152 (1971)
7.23   C. V. Shank, J. E. Bjorkholm, H. Kogelnik: Appl. Phys. Lett. **18**, 295 (1971)
7.24   J. E. Bjorkholm, C. V. Shank: Appl. Phys. Lett. **20**, 3 (1972)
7.25   J. E. Bjorkholm, C. V. Shank: IEEE J. QE-**8**, 833 (1972)
7.26   S. Chandra, N. Takendei, S. R. Hartmann: Appl. Phys. Lett. **21**, 144 (1972)
7.27   A. N. Rubinov, T. Sh. Efendiev: Sov. J. Quant. Electron. **3**, 268 (1973)
7.28   T. Sh. Efendiev, A. N. Rubinov: J. Appl. Spectrosc. **21**, 526 (1974)
7.29   T. Sh. Efendiev, A. N. Rubinov: Sov. J. Quantum Electron. **2**, 858 (1975)
7.30   J. S. Bakos, J. Füzessy, Zs. Sörlei, J. Szigeti: Phys. Lett. **50A**, 227 (1974)
7.31   A. N. Rubinov, T. Sh. Efendiev, A. V. Adamushko, J. Bor: Optics Commun. **18**, 18 (1976)
7.32   V. I. Vashchuk, K. F. Gorot', G. Yu. Kozak, N. N. Malykhina, E. A. Tikhonov: Sov. J. Quantum. Electron. **10**, 1006 (1981)
7.33   M. Sargent III, W. H. Swantiner, J. D. Thomas: IEEE J. QE-**16**, 465 (1980)
7.34   A. N. Rubinov, T. Sh. Efendiev: Sov. J. Quantum Electron. **12**, 1539 (1982)
7.35   Zs. Bor: Appl. Phys. **19**, 39 (1979)
7.36   Zs. Bor: Opt. Commun. **29**, 103 (1979)
7.37   Zs. Bor: IEEE J. QE-**16**, 517 (1980)
7.38   Zs. Bor, Alexander Müller, B. Racz, F. P. Schäfer: Appl. Phys. **B27**, 9 (1982)
7.39   Zs. Bor, Alexander Müller, B. Racz, F. P. Schäfer: Appl. Phys. **B27**, 77 (1982)
7.40   Zs. Bor, F. P. Schäfer: Appl. Phys. **B31**, 209 (1983)
7.41   M. Gottlieb, C. L. M. Ireland, J. M. Ley: "Electro-optic and acousto-optic scanning and deflection" in *Optical Engineering*, Vol. 3 (Dekker, New York 1983)
7.42   G. T. Sincerbox, G. Roosen: Appl. Opt. **22**, 690 (1983)
7.43   G. Roosen, M.-T. Plantegenest: Opt. Commun. **47**, 358 (1983)
7.44   J. P. Huignard, B. Ledu: Opt. Lett. **7**, 310 (1982)
7.45   M. P. Petrov, S. V. Miridonov, S. I. Stepanov, V. V. Kulikov: Opt. Commun. **31**, 301 (1979)
7.46   M. P. Petrov, S. I. Stepanov, A. A. Kamshilin: Opt. Commun. **29**, 44 (1979)
7.47   H. J. Gerritsen, E. G. Ramberg, S. Freeman: "Image processing with nonlinear optics" in *Proc. Symp. on Modern Optics* (Polytechnic Institute of Brooklyn, New York 1967) p. 109
7.48   D. M. Bloom, C. V. Shank, R. L. Fork, O. Teschke: In *Picosecond Phenomena*, ed. by C. V. Shank, E. P. Ippen, S. L. Shapiro, Springer Ser. Chem. Phys., Vol. 4 (Springer, Berlin, Heidelberg 1978) p. 372
7.49   A. Morimoto, T. Kobayashi, T. Sueta: Jap. Appl. Phys. **20**, 1129 (1981)
7.50   H. Vanherzeele, J. L. Van Eck: Appl. Opt. **20**, 524 (1981)
7.51   J. G. Fujimoto, E. P. Ippen: Opt. Lett. **8**, 446 (1983)
7.52   P. Günter: Phys. Rept. **93**, 199 (1982)
7.53   M. Z. Zha, P. Günter: Opt. Lett. **10**, 187 (1985)

7.54 P. Yeh: Optics Commun **45**, 323 (1983) and J. Opt. Soc. Am. **73**, 1268 (1983)
7.55 Y. H. Ya: Optics and Quantum Electronics **14**, 574 (1982)
7.56 D. M. Pepper, R. L. Abrams: Opt. Lett. **3**, 212 (1978)
7.57 J. Nilsen, A. Yariv: Appl. Phys. **18**, 143 (1979)
7.58 J. Nilsen, A. Yariv: J. Opt. Soc. Am. **71**, 180 (1981)
7.59 L. K. Lam, R. W. Hellwarth: "A wide-angle narrow-band optical filter using phase-conjugation by four-wave mixing in a waveguide" 11th Intern. Quant. Electr. Conf. Boston, MA (1980); Ref. 6 in [7.60]
7.60 J. Nilsen, N. S. Gluck, A. Yariv: Opt. Lett. **6**, 380 (1981)
7.61 J. Nilsen, A. Yariv: Opt. Commun. **39**, 199 (1981)
7.62 S. Saikan, H. Wakata: Opt. Lett. **6**, 281 (1981)
7.63 D. Veleskas, K. Jarashiunas, P. Baltrameiunas, J. Vaitkus: Pisma, J. Teor. Fiz. **1**, 708 (1975)
7.64 H. J. Eichler, U. Klein, D. Langhans: Appl. Phys. **21**, 215 (1980)
7.65 H. J. Eichler, G. Enterlein, D. Langhans: Appl. Phys. **23**, 299 (1980)
7.66 M. Born, E. Wolf: *Principles of Optics* (Pergamon, London 1970)
7.67 M. Maier, W. Kaiser, J. A. Giordmaine: Phys. Rev. Lett. **17**, 1275 (1966)
7.68 D. C. Champeney: *Fourier Transforms and their Physical Applications* (Academic, London 1973)
7.69 D. L. Bradley: In *Ultrashort Light Pulses*, ed. by S. L. Shapiro, Topics Appl. Phys., Vol. 18 (Springer, Berlin, Heidelberg 1977) Chap. 2
7.70 R. L. Fork, B. I. Greene, C. V. Shank: Appl. Phys. Lett. **38**, 671 (1981)
7.71 E. P. Ippen, C. V. Shank: In *Ultrashort Light Pulses*, ed. by S. L. Shapiro, Topics Appl. Phys., Vol. 18 (Springer, Berlin, Heidelberg 1977) Chap. 3
7.72 Ch. J. Kennedy, J. C. Matter, A. L. Smirl, H. Weiche, F. A. Hopf, S. V. Pappu: Phys. Rev. Lett. **32**, 419 (1974)
7.73 C. V. Shank, D. H. Auston: Phys. Rev. Lett. **34**, 479 (1975)
7.74 H. J. Eichler, U. Klein, D. Langhans: Appl. Phys. **21**, 215 (1980)
7.75 Z. Vardeny, J. Tauc: Opt. Commun. **39**, 396 (1981)
7.76 C. V. Shank, E. P. Ippen: Appl. Phys. Lett. **26**, 62 (1975)
7.77 B. Wilhelmi, J. Herrmann: Kvantovaja Elektronika **7**, 1876 (1980)
7.78 A. von Jena, H. E. Lessing: Appl. Phys. **19**, 131 (1979)
7.79 D. Reiser, A. Laubereau: Appl. Phys. **B27**, 115 (1982)
7.80 D. W. Phillion, D. J. Kuizenga, A. E. Siegmann: Appl. Phys. Lett. **27**, 85 (1975)
7.81 D. Langhans: Dissertation, Technische Universität Berlin (1980)
7.82 P. L. Liao, N. P. Economou, R. R. Freeman: Phys. Rev. Lett. **39**, 2473 (1977)
7.83 P. F. Liao, D. M. Bloom, N. P. Economou: Appl. Phys. Lett. **32**, 813 (1978)
7.84 J. P. Woerdman, M. F. H. Schuurmans: Opt. Lett. **6**, 239 (1981)
7.85 J. F. Lam, D. G. Steel, R. A. McFarlane, R. C. Lind: Appl. Phys. Lett. **38**, 977 (1981)
7.86 Y. Fukuda, K. Yamada, T. Hashi: J. Phys. Soc. Japan Lett. **48**, 1403 (1980)
7.87 Y. Fukuda, K. Yamada, T. Hashi: J. Phys. Soc. Japan **50**, 592 (1981)
7.88 Y. Fukuda, K. Yamada, T. Hashi: Optics Commun. **37**, 299 (1981)
7.89 M. Kroll: Opt. Lett. **7**, 151 (1982)
7.90 R. C. Lind, D. G. Steel, M. B. Klein, R. L. Abrams, C. R. Giuliano, R. K. Jain: Appl. Phys. Lett. **34**, 147 (1979)
7.91 R. A. Fisher, B. J. Feldman: Opt. Lett. **4**, 140 (1979)
7.92 D. G. Steel, R. C. Lind, J. F. Lam, C. R. Giuliano: Appl. Phys. Lett. **35**, 376 (1979)
7.93 P. Aubourg, J. P. Bettini, G. P. Agrawal, P. Cottin, D. Guerin, O. Meunier, J. L. Boulnois: Opt. Lett. **6**, 383 (1981)
7.94 G. P. Agrawal, A. van Lerberghe, P. Aubourg, J. L. Boulnois: Opt. Lett. **7**, 540 (1982)
7.95 D. Bloch, M. Ducloy: J. Opt. Soc. Am. **73**, 635 (1983)
7.96 J. F. Lam: Optical Engineering **21**, 219 (1982)
7.97 M. Ducloy: "Nonlinear optical phase conjugation" in *Festkörperprobleme*, Vol. 22 (Vieweg, Braunschweig 1982) pp. 35–60

Chapter 8

8.1   T. Tamir (ed.): *Integrated Optics*, 2nd ed., Topics Appl. Phys., Vol. 7 (Springer, Berlin, Heidelberg 1982)

8.2   R. G. Hunsperger: *Integrated Optics: Theory and Technology*, 2nd. ed., Springer Ser. Opt. Sci., Vol. 33 (Springer, Berlin, Heidelberg 1984)

8.3   D. Casasent (ed.): *Optical Data Processing*, Topics Appl. Phys., Vol. 23 (Springer, Berlin, Heidelberg 1978)

8.4   S. H. Lee (ed.): *Optical Information Processing*, Topics Appl. Phys., Vol. 48 (Springer, Berlin, Heidelberg 1981)

8.5   H. J. Nussbaumer: *Fast Fourier Transform and Convolution Algorithm*, 2nd. ed., Springer Ser. Inf. Sci., Vol. 2 (Springer, Berlin, Heidelberg 1982)

8.6   T. S. Huang (ed.): *Two Dimensional Digital Signal Processing I & II*, Topics Appl. Phys., Vol. 42 and 43 (Springer, Berlin, Heidelberg 1981)

8.7   G. Mourou, D. M. Bloom, C.-H. Lee (eds.): *Picosecond Electronics and Optoelectronics*, Springer Ser. Electrophys., Vol. 21 (Springer, Berlin, Heidelberg 1985)

Additional References

Interest in dynamic gratings and their applications has grown rapidly in the last few years. A large number of papers have been published recently which are only partially included in the references given above. A survey of the latest developments will be published by the *IEEE Journal of Quantum Electronics* in a *special issue* on *Dynamic Gratings and Four-wave mixing*, which is scheduled to appear in *August 1986* or in the following months. The main topics to be covered are listed below. This issue will contain more than 500 relevant references included in the following papers:

Four-wave Mixing and Grating Theory

Enns, R.H.: Inverse scattering and the three-wave interaction in nonlinear optics
Fujimoto, J.F., Yee, T.K.: Diagrammatic density matrix theory of transient four-wave mixing and the measurement of transient phenomena
Shen, Y.R.: Basic considerations of four-wave mixing and dynamic gratings

Gases, Liquids, Microemulsions, Dye Solutions

Freysz, E., Claeys, W., Ducasse, A.: Dynamic gratings induced by electrostrictive compression of critical microemulsions
Le Boiteux, S., Simoneau, P., Bloch, D., De Oliveira, F.A.M., Ducloy, M.: Saturation behaviour of resonant degenerate four-wave and multiwave mixing in the Doppler-broadened regime: experimental analysis of a low-pressure Ne discharge
Scott, A.M., Hazell, M.S.: High efficiency scattering in transient Brillouin enhanced four-wave mixing
Todorov, T., Nikolova, L., Tomova, N., Dragostinova, V.: Photo-induced anisotropy in rigid dye solutions for transient holography

Liquid Crystals

Arakelian, S.M., Chilingarian, Ju.S.: Dynamic self-diffraction effects in liquid crystals
Khoo, I.C.: Dynamic gratings and the associated self-diffraction and wave front conjugation processes in liquid crystals
Madden, P.A., Saunders, F.C., Scott, A.M.: Degenerate FWM in the isotropic phase of liquid crystals: the influence of molecular structure

Semiconductors

Aoyagi, Y., Segawa, Y., Namba, S.: Study of the dynamics of excited states in CdS and CuCl by transient grating techniques
Baumert, R., Broser, I., Buschnik, K.: Nonlinearity in the refractive index due to an excitonic molecular resonance state in CdS

Bergner, H., Brückner, V., Supianek, M.: Ultrafast processes in silicon studied by transient gratings

Jarasiunas, K., Stomp, S., Sirmulis, E.: Transient gratings in InSb at two-photon excitation

Kalt, H., Renner, R., Klingshirn, C.: Resonant self-diffraction from dynamic laser-induced gratings in II–VI compounds

McKenzie, D.A., Hagan, D.J., Al-Attar, H.A.: Four-wave mixing in InSb

Shiren, N.S., Melcher, R.L., Kazyaka, T.G.: Multiple quantum phase conjugation in microwave acoustics

Vaitkus, J., Jarasiunas, K., Gaubas, E., Jonikas, L.: Diffraction of light by transient gratings in crystalline, ion-implanted and amorphous silicon

FWM in Guided Wave and Injection Laser Structures

Nakajima, H., Frey, R.: Collinear nearly degenerate four-wave mixing in intracavity amplifying media

Stegeman, G.I., Seaton, C.T., Karaguleff, C.: Degenerate four-wave mixing with guided waves

Inorganic Crystals: Electronic Energy Migration

Morgan, G.P., Chen, S.Z., Yen, W.M.: Transient grating spectroscopy of $LaP_5O_{14}:Nd^{3+}$

Powell, R.C., Tyminski, J.K., Ghazzawi, A.M., Lawson, C.M.: Dynamics of population gratings in NdP_5O_{14}

Photorefractive Materials

Carrascosa, M., Agullo-Lopez, F.: Kinetics of optical erasure of sinusoidal holographic gratings in photorefractive materials

Jonathan, J.M.C., Hellwarth, R.W., Roosen, G.: Effect of applied electric fields on the buildup and decay of photorefractive materials

Miteva, M.G.: Some possibilities for improving the holographic recording characteristics in $Bi_{12}SiO_{20}$ monocrystals

Ringhofer, K.H., Rupp, R.A.: Light-induced scattering in photorefractive materials

Surface Gratings and Solid State Phase Changes

Marine, W., Mathiez, P.: Dynamics of laser annealing of amorphous Ge and GaAs films by the transient grating method

Siegman, A.E., Fauchet, P.M.: Stimulated Wood's anomalies on laser-illuminated surfaces

Ultrafast Phenomena

Farrar, M.R., Cheng, Lap-Tak, Yan, Young-Xiu, Nelson, K.A.: Impulsive stimulated Brillouin scattering in KD_2PO_4 near the structural phase transition

Fayer, D.: Picosecond holographic grating generation of ultrasonic waves

Iadutilin, V.S.: Dynamic holography for ultrashort light pulses

Siegman, A.E., Fauchet, P.M.: Stimulated Wood's anomalies on laser-illuminated surfaces

Trebino, R., Barker, C.E., Siegman, A.E.: Tunable laser-induced gratings for the measurement of ultrafast phenomena

Vaitkus, J., Jarasiunas, K., Gaubas, E., Jonikas, L.: Diffraction of light by transient gratings in crystalline, ion-implanted and amorphous silicon

Hydrodynamics and Ultrasonic Investigations

Charmet, J.C., Cloitre, M., Fermigier, M., Guyon, E., Jenffer, P., Limat, L., Petit, L.: Application of forced Rayleigh scattering to hydrodynamic measurements
Farrar, M.R., Cheng, Lap-Tak, Yan, Youg-Xiu, Nelson, K.A.: Impulsive stimulated Brillouin scattering in KD_2PO_4 near the structural phase transition
Fayer, D.: Picosecond holographic grating generation of ultrasonic waves
Shiren, N.S., Melcher, R.L., Karzyaka, T.G.: Multiple quantum phase conjugation in microwave acoustics

Photochemistry

Burland, D.M.: Holographic methods for investigating solid state photochemistry
Deeg, F.W., Pinsel, J., Bräuchle, Chr.: New grating experiments in the study of irreversible photochemical reactions
Hochstrasser, R.M., Myers, A.B.: Comparison of four-wave mixing techniques for studying orientational relaxation

Coherent Beam Amplification

Tschudi, T., Herden, A., Goltz, J., Klumb, H., Laeri, F.: Image amplification by two- und four-wave mixing in $BaTiO_3$ photorefractive crystals
Vinetskii, V.L., Levskin, A.E., Tomschik, P.M., Chumnak, A.A.: Theory of dynamic transformation of light beams by conduction electrons in semiconductors

Oscillators with Dynamic Grating Feedback

Bor, Z., Müller, A.: Picosecond distributed feedback dye lasers
Kwong, S.K., Cronin-Golomb, M., Yariv, A.: Oscillation with photorefractive gain
Onhayoun, M., Guern, Y.: Laser mirror by degenerate four-wave mixing in a saturable absorber

Alternative and Related Technicques

Baumert, R., Broser, I., Buschnik, K.: Nonlinearity in the refractive index due to an excitonic molecular resonance state in CdS
Dorion, P., Lalanne, J.R., Pouligny, B.: Spatial Fourier spectrum analysis of an induced thermal lens: a complementary method to thermal dynamic gratings
Shiren, N.S., Melcher, R.L., Kazyaka, T.G.: Multiple quantum phase conjugation in microwave acoustics

Subject Index

Inverse Source Problems

in Optics

Editor: **H. P. Baltes**

With a foreword by J.-F. Moser

1978. 32 figures. XI, 204 pages. (Topics in Current
Physics, Volume 9). ISBN 3-540-09021-5

Contents: *H. P. Baltes:* Introduction. –
H. A. Ferwerda: The Phase Reconstruction Problem
for Wave Amplitudes and Coherence Functions. –
B. J. Hoenders: The Uniqueness of Inverse
Problems. – *H. G. Schmidt-Weinmar:* Spatial Resolu-
tion of Subwavelength Sources from Optical Far-
Zone Data. – *H. P. Baltes, J. Geist, A. Walther:* Ra-
diometry and Coherence. – *A. Zardecki:* Statistical
Features of Phase Screens from Scattering Data.

Inverse Scattering Problems

in Optics

Editor: H. P. Baltes

With a Foreword by R. Jost

1980. 49 figures, 2 tables. XIV, 313 pages. (Topics
in Current Physics, Volume 20)
ISBN 3-540-10104-7

Contents: *H. P. Baltes:* Progress in Inverse Optical
Problems. – *G. Ross, M. A. Fiddy, M. Nieto-Vespe-
rinas:* The Inverse Scattering Problem in Structural
Determinations. – *E. Jakeman, P. N. Pusey:* Photon-
Counting Statistics of Optical Scintillation. –
A. Selloni: Microscopic Models of Photodetection. –
M. Bertero, C. De Mol, G. A. Viano: The Stability of
Inverse Problems. – *R. Goulard, P. J. Emmerman:*
Combustion Diagnostics by Multiangular Absorp-
tion. – *W.-M. Boerner:* Polarization Utilization in
Electromagnetic Inverse Scattering.

Springer-Verlag
Berlin Heidelberg
New York Tokyo

Springer

B.Y.Zel'dovich. N.F.Pilipetsky, V.V.Shkunov

Principles of Phase Conjugation

1985. 70 figures. X, 250 pages. (Springer Series in Optical Sciences, Volume 42). ISBN 3-540-13458-1

Contents: Introduction to Optical Phase Conjugation. – Physics of Stimulated Scattering. – Properties of Speckle-Inhomogeneous Fields. – OPC by Backward Stimulated Scattering. – Specific Features of OPC-SS. – OPC in Four-Wave Mixing. – Nonlinear Mechanisms for FWM. – Other Methods of OPC. – References. – Subject Index.

Holographic Recording Materials

Editor: **H.M.Smith**
1977. 96 figures, 17 tables. XIII, 252 pages. (Topics in Applied Physics, Volume 20). ISBN 3-540-08293-X

Contents: *H.M.Smith:* Basic Holographic Principles. – *K.Biedermann:* Silver Halide Photographic Materials. – *D.Meyerhofer:* Dichromated Gelatin. – *D.L.Staebler:* Ferroelectric Crystals. – *R.C.Duncan Jr., D.L.Staebler:* Inorganic Photochromic Materials. – *J.C.Urbach:* Thermoplastic Hologram Recording. – *R.A.Bartolini:* Photoresists. – *J.Bordogna:* Other Materials and Devices.

V.M.Fridkin

Photoferroelectrics

1979. 63 figures, 3 tables. X, 174 pages. (Springer Series in Solid-State Sciences, Volume 9). ISBN 3-540-09418-0

Contents: Introduction. – The Thermodynamics of Photoferroelectrics. – The Microscopic Theory of Photoferroelectric Phenomena. – Screening of Spontaneous Polarization. – Photoferroelectric Phenomena and Photostimulated Phase Transitions. – The Anomalous Photovoltaic Effect in Ferroelectrics. – The Photorefractive Effect in Ferroelectrics. – Screening Phenomena. – References. – Subject Index.

Springer-Verlag
Berlin Heidelberg
New York Tokyo

Springer

Springer Series in Optical Sciences

Editorial Board: J.M. Enoch D.L. MacAdam A.L. Schawlow K. Shimoda T. Tamir